Revise AS Biology for AQA Specification A

Graham Read, Ray Skwierczynski and Simon Burch

Heinemann

Contents

Introduction – How to use this revision guide

This revision guide is for the AQA Biology AS course, Specification A. It is intended for pupils taking either 'Biology' or 'Human Biology'. It is divided into three modules, details of which are given on the next page.

Pupils taking the 'Biology' course should use the pages displaying this symbol...   **B**

Pupils taking 'Human Biology' should follow the pages displaying this symbol...   **HB**

Whichever course you are following, you will be taking tests – either at the end of each module or all at the end of the course.

Each module begins with an **introduction**, which summarises the content. It also reminds you of the topics from your GCSE course which the module draws on.

The content of each module is presented in **blocks**, to help you divide up your study into manageable chunks. Each block is dealt with in several spreads. These do the following:

- summarise the **content**;
- indicate **points to note**;
- include **diagrams** of the sort you might need to label in tests;
- provide **quick check** questions to help you test your understanding. A box in the text, like the one shown here, indicates when you should be able to attempt a particular question.

 ✓ *Quick check 3*

At the end of each module, there are longer **end-of-module questions** similar in style to those you will encounter in tests. **Answers** to all questions are provided at the end of the book.

You need to understand the **scheme of assessment** for your course. This is summarised on the next page. At the end of the book you will find some **exam tips** to help you prepare for the examinable component of the course.

AQA AS Biology – Assessment

You can take an AS (Advanced Subsidiary) as a qualification on its own or as the first part of an A level (Advanced Level) qualification. An AS forms 50% of an Advanced Level.

The AS offered by AQA Specification (Syllabus) A offers two alternatives: 'Biology' or 'Human Biology'.

- **Module 1: Molecules, Cells and Systems** is common to both courses.
- **Module 2: Making Use of Biology** is for those taking AS Biology.
- **Module 3: Pathogens and Disease** is for those taking AS Human Biology.

Note that there are some topics common to both Module 2 and Module 3 – **Cell division**, **The genetic code** and **Gene technology**. In this book, these common topics are included in Module 2 only (to avoid repetition). Students taking Human Biology will need to refer to **pages 36–47** when revising Module 3.

- **Module 4** (centre-assessed coursework) is dealt with by your school or college.

This book is designed to prepare you for the exams for Modules 1–3.

About the tests

To get an AS, you will have to take a 1½ hour exam for each Module. All the questions have to be answered in each exam. Most of the marks will be for remembering and being able to explain the information in each Module. What you need to know is given in this book, at the level you need to know it. If you know more than that, it certainly will not harm you – but it will not be needed to pass an exam. There is no synoptic element in these exams (that comes at Advanced Level). This means that you do not have to know the information in Molecules, Cells and Systems to pass the other exam.

The table shows an outline of the AQA Specification A for AS.

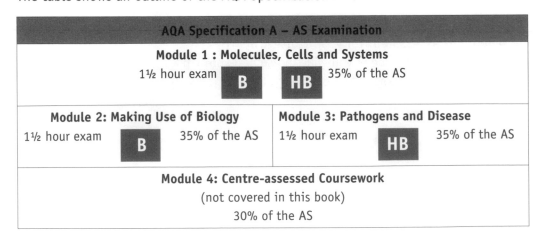

AQA Specification A – AS Examination		
Module 1 : Molecules, Cells and Systems		
1½ hour exam **B** **HB** 35% of the AS		
Module 2: Making Use of Biology		**Module 3: Pathogens and Disease**
1½ hour exam **B** 35% of the AS		1½ hour exam **HB** 35% of the AS
Module 4: Centre-assessed Coursework (not covered in this book) 30% of the AS		

Module 1: Molecules, Cells and Systems

This module is broken down into seven topics: Cells and their structure; Plasma membranes; Biological molecules; Enzymes; Tissues and organs; Blood system and breathing; and The heart.

Cells and their structure

- Life exists as cells; there are two types: prokaryotic and eukaryotic.
- Light microscopes magnify enough to make cells visible but lack the greater resolution and effective magnification of electron microscopes used to see details of cellular structure.

Plasma membranes

- The cell surface membrane is differentially/selectively permeable and controls what enters and leaves the cell.
- Cells exchange substances with their environment by diffusion, osmosis, active transport and endo/exocytosis.

Biological molecules

- Organisms are made up of carbon-based biological molecules.
- Small monomer molecules can join together to form very large polymers.
- Other elements attached to carbon chains or rings give more possibilities.
- Biological molecules are grouped into types with similar properties: carbohydrates, proteins, lipids and nucleic acids (dealt with in Module 2).

Enzymes

- Enzymes are proteins that are biological catalysts.
- They allow reactions to take place quickly in environmental conditions found in cells.

Tissues and organs

- Large organisms have special exchange surfaces and transport systems to carry substances to and from their tissues.
- Exchange of substances with the environment and cells still involve diffusion, osmosis and active transport across cell membranes.

Blood system and breathing

- In humans, blood flows in blood vessels, pushed by pumping of the heart.
- Oxygen is carried from the exchange surfaces of the lungs to the tissues and waste carbon dioxide is carried back to the lungs to be excreted.
- Exchange of substances between blood and tissues occurs in capillaries.

The heart

- During exercise muscles use more energy from increased respiration.
- Heartbeat and breathing rates increase to supply more oxygen and respiratory substrates.
- These responses involve receptors in blood vessels and the medulla in the brain.

B
HB

Cells and microscopy

The cell is the basic unit of living organisms. Cells come only from existing cells by cell division. Cells and organisms can be divided into two main groups: **prokaryotes** and **eukaryotes**. Microscopy (the use of microscopes) has proved invaluable in investigating the structure of cells.

Comparison of prokaryotes and eukaryotes

Bacteria are prokaryotes and are very small single-celled organisms with no nucleus and no other membrane-bound organelles.

● A typical bacterial cell will have a cell wall, cell membrane, genetic material (DNA), small (70S) ribosomes and cytoplasm.

● A capsule, plasmids and flagellae may be present.

Eukaryotes form the vast majority of types of living organisms, including plant and animal. Eukaryotic cells have a nucleus and many other membrane-bound organelles.

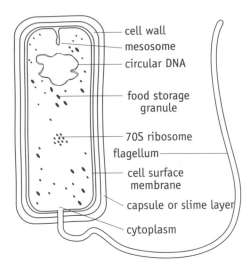

Typical bacterial cell as seen with an electron microscope

Prokaryotic	Eukaryotic
DNA is circular, not in a nucleus	DNA is linear, within a nucleus
Diameter of cell 0.5–10 µm	Diameter of cell 10–100 µm
Smaller, lighter 70S ribosomes	Larger, heavier 80S ribosomes
No mitochondria present	Mitochondria present
No Golgi body	Golgi body present
Flagella (when present) simple, lacking microtubules	Flagella (when present) have microtubules

✓ *Quick check 1*

Eukaryotic plant cells also have:

● cellulose cell walls providing support and shape to the cell;

● starch grains;

● chloroplasts (in photosynthetic cells), containing the pigment chlorophyll;

● a large vacuole containing soluble sugars, salts and sometimes pigment.

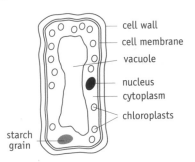

A palisade mesophyll cell as seen with a light microscope

Light microscopy

Light microscopy reveals the position of the cell membrane, cytoplasm and nucleus in eukaryotic cells. Small organelles cannot be seen clearly. In light microscopy:

● light passes through a specimen and lenses magnify the image;

● the resolving power (the ability to distinguish close objects) is limited by the wavelength of light;

● the maximum useful magnification is × 1500.

The total magnification is calculated by multiplying the magnifications of the eyepiece lens and objective lens, e.g. × 10 and × 40 gives a total magnification of × 400.

Micrometry – estimation of size

The size of structures can be measured using microscopy and specially designed scales – micrometers.

- One micrometer is placed in the eyepiece and one on the stage.
- The stage micrometer has known divisions, e.g. 0.01 mm.
- At each magnification, the number of eyepiece divisions corresponding to a particular distance on the stage micrometer is recorded.
- A structure can then be measured in eyepiece units and its size calculated.

Electron microscopy

The electron microscope is used to investigate the fine structure (**ultrastructure**) of a cell. It uses a beam of electrons focused by electromagnets, as opposed to light rays focused by lenses in light microscopy.

Advantages of electron microscopy

- Electrons have a shorter wavelength than light, giving **greater resolution** (the ability to distinguish between two close objects). Light microscopy has a maximum resolution of around 200 nm, with the electron microscope it is 0.5 nm. The **maximum useful magnification** is higher than with light microscopy.

Disadvantages of electron microscopy

- As a vacuum is required, living specimens cannot be seen.
- Preparation and staining techniques can alter/damage cells.
- It is expensive and expert training is required in its use.

total length = 1 mm

0 10 20 30 40 50 60 70 80 90 100

stage micrometer in centre of glass slide

2 divisions on eyepiece micrometer equivalent to 0.45 mm

0 1 2 3 4 5 6 7 8 9 10

eyepiece micrometer

Micrometry

✓ **Quick check 2**

Types of electron microscope

The **transmission electron microscope** (TEM):
- uses electromagnets to focus an electron beam, which is transmitted through the specimen;
- results in denser parts of a specimen appearing darker, as they absorb more electrons (contrast is improved using electron-dense stains);
- provides a high resolution (0.5 nm), but only thin sections can be examined.

The **scanning electron microscope** (SEM):
- scans an electron beam across the specimen;
- collects reflected electrons in a cathode ray tube to form a TV image;
- shows the surface of a structure, has a greater depth of field and can provide 3-D images;
- has a lower resolution (5–20 nm) than the TEM.

Do not confuse resolution and magnification. You could get a magnification of 500 000 with a light microscope but the image would be very blurred due to poor resolution.

? Quick check questions

1 Give three ways in which prokaryotes and eukaryotes differ in structure.
2 Give one advantage and one disadvantage of using an electron microscope.

B
HB

Cell ultrastructure – organelles

Organelles in eukaryotic cells include the nucleus, mitochondria, Golgi body, lysosomes, rough and smooth endoplasmic reticulum, chloroplasts and ribosomes.

> ▶ You must be able to recognise the different cell organelles as shown below.

Nucleus

The nucleus contains the genetic material, DNA, and:

- is bound by a double membrane, the **nuclear envelope**, which has **nuclear pores** allowing communication with the cytoplasm;

- contains **chromatin** (DNA and protein), and **nucleoli** (in a dividing cell the chromatin is seen as **chromosomes**);

- controls protein synthesis, cell division and the production of ribosomes.

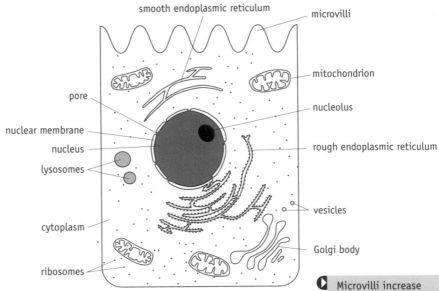

An epithelial cell as seen with an electron microscope

> ▶ Microvilli increase the surface area of a cell for the absorption of substances. The term 'brush border' is sometimes used to describe microvilli across the top of a cell.

Mitochondria

Mitochondria are involved in **aerobic respiration** that produces ATP. They are often rod shaped and 1–10 μm in length. Mitochondria:

- are bounded by two membranes forming an envelope around the **matrix** – the site of enzymes involved in respiration and also containing DNA and ribosomes;

- have an inter-membrane space between the outer and the inner membranes;

- have **cristae** – folds of the inner membrane which provide a large surface area for enzymes associated with ATP production needing oxygen (oxidative phosphorylation).

Cells requiring large amounts of ATP have numerous mitochondria, e.g. muscle cells and those involved in active transport.

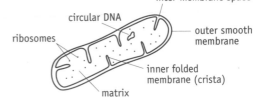

A mitochondrion

Ribosomes

Ribosomes are very small organelles (20 nm diameter), made of one large and one small sub-unit, each made up of protein and RNA. Ribosomes:

- can be present in the cytoplasm singly, in a chain (polysomes) or attached to the rough endoplasmic reticulum;

- are used in **protein synthesis** – where condensation reactions join amino acids together.

Endoplasmic reticulum

This consists of flattened membrane sacs, called **cisternae**.

- The surface of **rough endoplasmic reticulum** (RER) has ribosomes that produce proteins in the cisternae.
- **Smooth endoplasmic reticulum** lacks ribosomes and is involved in the transport of substances around the cell, and the production of lipids.

Golgi body

A Golgi body consists of a stack of flattened, membrane-bound sacs – **cisternae**.

- The cisternae are continually being formed at one end and pinched off as **Golgi vesicles** at the other.
- Secretory cells have large Golgi bodies.
- Golgi bodies produce glycoproteins, package and secrete enzymes, form cell wall material in plant cells, and form lysosomes.

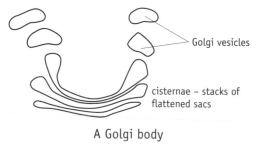

Golgi vesicles

cisternae – stacks of flattened sacs

A Golgi body

Lysosomes

Lysosomes consist of membrane-bound sacs containing digestive enzymes, e.g. lipases and proteases. Lysosomes are used in:

- some microorganisms (e.g. Amoeba) to digest food;
- phagocytic white blood cells to destroy bacteria;
- the breakdown of cells during development;
- damaged cells for self-digestion.

> ❶ It is quite easy to learn to label the parts of these organelles – make sure you do!

Chloroplasts

Chloroplasts are found only in photosynthetic plant cells.

- They are flattened discs, 3–10 μm in diameter, bounded by two membranes.
- Inside is a membrane system of many flattened sacs called **thylakoids**, which in places form stacks called **grana**.
- Grana contain chlorophyll molecules that **absorb light** for **photosynthesis**.
- The membrane system is surrounded by the **stroma** – the site of the enzymes used to make sugars and starch during photosynthesis.

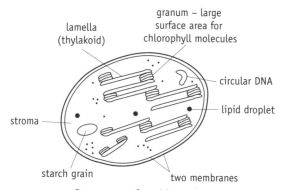

lamella (thylakoid)

granum – large surface area for chlorophyll molecules

circular DNA

lipid droplet

stroma

starch grain

two membranes

Structure of a chloroplast

✓ Quick check 1, 2

❓ Quick check questions

1 Give two ways in which the structure of a mitochondrion and a chloroplast are similar.

2 Give one function of the following organelles:
   (a) ribosomes; (b) smooth endoplasmic reticulum; (c) Golgi body.

B
HB

Cell fractionation

Cell fractionation refers to the break-up of cells and the centrifugation of the resulting suspension, so that different organelles can be isolated. In order to separate all the organelles, ultracentrifugation (high speed centrifugation) is required.

Method of cell fractionation

Centrifugation separates structures of different mass/density and size. Differential centrifugation involves centrifuging at different speeds to separate and isolate the different organelles in a cell.

- Cells are broken open and **homogenised** by grinding a tissue such as liver in ice-cold, **isotonic** buffer solution, using a blender.

- An isotonic solution prevents the net osmotic movement of water in or out of organelles, which might cause them to burst or shrivel.

- A low temperature prevents the action of enzymes that might cause self-digestion (autolysis) of organelles.

- The suspension is poured into a tube and spun in a centrifuge at low speed. Cell debris, e.g. cell walls of plant cells, forms a **sediment** (**pellet**) at the bottom.

- The **supernatant** (liquid above the sediment) contains suspended organelles and is spun at a higher speed (generating greater force).

- The heaviest organelles, the **nuclei**, are forced to the bottom and form a sediment, whilst lighter organelles remain suspended in the supernatant.

- The procedure is repeated, increasing the speed of centrifugation each time, giving a series of pellets containing organelles of decreasing density.

- The organelles are usually isolated in the order: nuclei, chloroplasts, mitochondria, endoplasmic reticulum, lysosomes, and finally ribosomes.

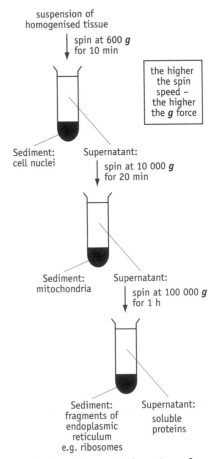

suspension of
homogenised tissue

spin at 600 *g*
for 10 min

the higher
the spin
speed –
the higher
the *g* force

Sediment:
cell nuclei

Supernatant:

spin at 10 000 *g*
for 20 min

Sediment:
mitochondria

Supernatant:

spin at 100 000 *g*
for 1 h

Sediment:
fragments of
endoplasmic
reticulum
e.g. ribosomes

Supernatant:
soluble
proteins

Differential centrifugation of
cell organelles

✓ *Quick check 1*

✓ *Quick check 2*

Isotonic refers to
solutions with the
same water
potential but not
necessarily the
same composition.

? *Quick check questions*

1 Explain why (**a**) a low temperature and (**b**) an isotonic solution are required during centrifugation.

2 Which will be the main organelle present in the first sediment during differential centrifugation?

The plasma membrane

The entry and exit of substances to cells are controlled by the **cell surface membrane** or **plasma membrane**, which surrounds the cytoplasm of a cell. The structure of the plasma membrane is described by the **fluid-mosaic model** and is basically the same for membranes around cell organelles.

The fluid-mosaic model

The cell membrane consists of protein and phospholipid. The **phospholipid** molecules form a double layer (**bilayer**) but are constantly moving about, giving a **fluid** structure. The protein molecules are unevenly distributed throughout the membrane, forming a **mosaic**. The **selective permeability** of the cell membrane is related to the type and distribution of protein and phospholipid molecules present in the membrane.

In the cell membrane:

- the **hydrophobic tails** (fatty acid chains) of phospholipid molecules are attracted towards each other, and the **hydrophilic heads** are orientated either inwards towards the cytoplasm or outwards towards the watery extra-cellular fluid, forming the phospholipid bilayer;
- protein molecules are embedded in the phospholipid bilayer;
- **cholesterol** decreases permeability and increases stability of the membrane;
- the phospholipid bilayer allows lipid-soluble molecules to pass through but restricts the passage of ions and polar molecules; these can pass through **channel proteins** (pores) that span the membrane;
- other protein molecules act as **carriers**, aiding the transport of ions and polar molecules (e.g. glucose) by facilitated diffusion and active transport;
- other protein molecules act as specific **receptors** for hormones (e.g. insulin), which attach to them and so allow the cell to respond;
- **glycoproteins**, composed of carbohydrate and protein, are on the outer surface of the membrane and are important in cell recognition, sometimes acting as **antigens**.

The proteins acting as channels, carriers and receptors are not all the same things!

✓ *Quick check 1, 2*

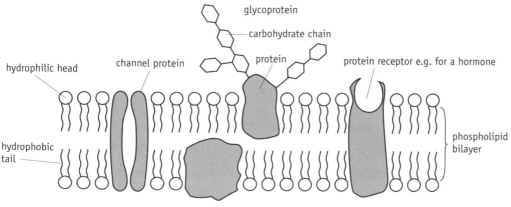

Structure of the plasma membrane

Quick check questions

1 Explain how lipid-soluble molecules are able to pass rapidly into a cell.
2 Give the function of: **(a)** protein carriers; **(b)** receptors in the cell membrane.

B / HB Transport across membranes

Movement of substances into and out of cells can occur by **diffusion, facilitated diffusion, osmosis, active transport, endocytosis** and **exocytosis**.

Diffusion

Diffusion is a passive process – it does not require energy from respiration.

> **Diffusion is the net movement of molecules from a higher concentration to a lower concentration until they are equally distributed.**

The rate of diffusion is increased by a greater concentration gradient, a large surface area (e.g. microvilli) and a short diffusion distance as indicated by **Fick's law**.

> **Rate of diffusion is proportional to** $\dfrac{\text{surface area} \times \text{difference in concentration}}{\text{thickness of exchange surfaces}}$

Facilitated diffusion

In this process channel proteins and carrier proteins transport ions and polar molecules across membranes. The specific tertiary structure of each type of protein means that it recognises, binds with and transports a specific substance.

- **Channel proteins** have a fixed shape whereas **carrier proteins** change shape as they move molecules, e.g. glucose, across the membrane.

- Facilitated diffusion does not require energy from respiration.

Water potentials and osmosis

Water moves from a place with a **less negative** (higher) **water potential to** a place with a **more negative** (lower) **water potential**.

- The water potential of **pure water is 0** (zero).

- **Solutions** have **negative water potentials** – the more concentrated the solution, the more negative the water potential.

- Water diffuses from a solution with a less negative water potential (hypotonic) into one with a more negative water potential (hypertonic) until the solutions are the same concentration (isotonic).

Osmosis is a special case of the **diffusion of water**.

> **Osmosis is the net movement of water molecules by diffusion from a solution of less negative water potential to a solution of more negative water potential through a partially permeable membrane.**

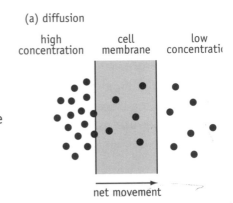

(a) diffusion

high concentration | cell membrane | low concentration

net movement

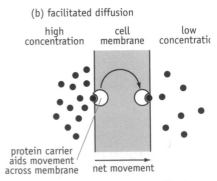

(b) facilitated diffusion

high concentration | cell membrane | low concentration

protein carrier aids movement across membrane

net movement

Diffusion and facilitated diffusion

> Get used to using the terms used here to describe differences in water potentials. References to larger and smaller or bigger and smaller water potentials will be marked wrong.

- The **cell surface membrane** is partially (and selectively) permeable.

- The membrane around the vacuole in plant cells is also partially permeable.

- Ions (and sugars) in the vacuole produce a negative water potential and water enters the vacuole by osmosis.

- This makes the vacuole (and cell) expand and push against the cell wall – making the cell **turgid** (firm).

- **Cell turgor** is essential in providing support in many plants.

selectively permeable membrane s.p.m.	
higher water potential	lower water potential
solution A 10% salt solution hypotonic to B	solution B 30% salt solution hypertonic to A

→ net osmotic movement of water

The process of osmosis

> In any question about movement of water in and out of cells it is very important to mention the net diffusion of water, osmosis, selectively permeable membranes and water potentials.

B

HB

✓ *Quick check 1, 2*

Active transport

Active transport is the transport of molecules or ions across a membrane by carrier proteins against a concentration gradient.

- It **requires energy** from respiration.

- Factors reducing respiration, reduce active transport:

    lower temperature

    lack of oxygen

    metabolic and respiratory inhibitors.

- Active transport involves carrier proteins in the membrane.

- The hydrolysis of ATP releases the energy required for active transport.

- Cells involved in active transport have a **large number of mitochondria** to provide the ATP required via **aerobic respiration**.

low concentration of molecules (e.g. glucose)

low concentration of molecules

protein carrier with specific shape for molecule to fit into

cell membrane

ATP used to make carrier change shape – carries molecule across membrane and releases it

ATP

ADP

high concentration of molecules

Active transport

Ψ = –230 kPa

Ψ = –150 kPa

Ψ = –190 kPa

Plant cells with different water potentials (arrows indicate the movement of water by osmosis)

> You can think of turgor as being like inflating the inner-tube of a bicycle tyre to make the tyre firm. The inner-tube is inflated with air and pushes against the rigid tyre – which resists expansion.

Endocytosis and exocytosis

- **Endocytosis** is the transport of large particles into the cell in vesicles formed by invagination of the cell surface membrane.

- **Exocytosis** is the reverse process and is used to secrete proteins, e.g. digestive enzymes, out of cells.

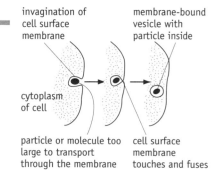

invagination of cell surface membrane

membrane-bound vesicle with particle inside

cytoplasm of cell

particle or molecule too large to transport through the membrane

cell surface membrane touches and fuses

✓ *Quick check 3, 4*

? Quick check questions

1 What is the water potential value of pure water?

2 Explain in terms of water potential the movement of water from the soil into a root hair of a plant cell.

3 Give two ways in which diffusion and active transport differ.

4 Give two ways by which you could reduce active uptake of ions by a cell.

Carbohydrates – mono- and disaccharides

Biological molecules such as carbohydrates, proteins and lipids are based on a small number of chemical elements. Carbohydrates contain the elements carbon, hydrogen and oxygen. The hydrogen and oxygen are present in the ratio of 2:1.

Carbohydrates can be classified into three groups: monosaccharides, disaccharides and polysaccharides. Monosaccharides and discaccharides are small, soluble molecules easy to transport and sweet to taste.

Monosaccharides (single sugars)

Monosaccharides are the basic molecular units (**monomers**) of carbohydrates. They are mainly used in respiration and in growth during the formation of larger carbohydrates. They:

- include **glucose**, which has the formula $C_6H_{12}O_6$ (**hexose**), but the elements can be arranged into different structures (**isomers**);

- are **reducing sugars** so give a positive (+ve) Benedict's test result (**brick red precipitate/colour**).

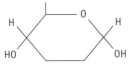

α glucose – used to make starch and glycogen

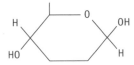

β glucose – used to make cellulose

Isomers of glucose

Disaccharides (double sugars)

A disaccharide is formed when two monosaccharides (**monomers**) are joined together by a **condensation reaction**. A **glycosidic bond** is formed between the monosaccharides.
An example of a disaccharide is **maltose**.

Glucose	+	glucose	→	maltose	+	water
$C_6H_{12}O_6$		$C_6H_{12}O_6$		$C_{12}H_{22}O_{11}$		H_2O
	monosaccharides			disaccharide		

✓ *Quick check 1, 2*

A disaccharide can be broken down into its monosaccharides by a **hydrolysis reaction**. Hydrolysis is use of water (hydro) in the breakdown (lysis) of a larger molecule into smaller molecules. Disaccharides can be **hydrolysed** by boiling with acid, e.g. dilute HCl, or by an enzyme at its **optimum** temperature.

Example: Hydrolysis of maltose

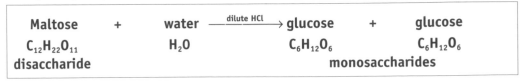

Maltose	+	water	dilute HCl →	glucose	+	glucose
$C_{12}H_{22}O_{11}$		H_2O		$C_6H_{12}O_6$		$C_6H_{12}O_6$
disaccharide					monosaccharides	

All the large biological molecules you study are formed from smaller molecules by condensation reactions – where water is formed. They can all be broken down by hydrolysis reactions – using water.

Benedict's test for reducing sugars

- A small amount of the food sample/solution being tested for reducing sugar is placed in a test tube with 2 cm^3 of **Benedict's solution**.

- This is heated in a boiling water bath for 5 minutes.

- A **brick red** precipitate/colour is a **positive** result.

- Glucose, fructose and maltose give positive results.

- If the Benedict's solution remains **blue – no reducing sugar** is present.

Test for a non-reducing sugar (sucrose)

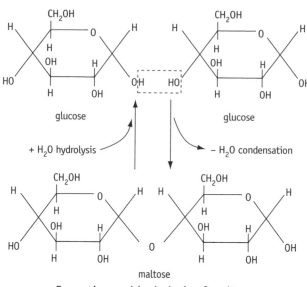

glucose   glucose

+ H$_2$O hydrolysis   – H$_2$O condensation

maltose

Formation and hydrolysis of maltose

Sucrose is a non-reducing sugar and can be identified by the following method.

- Carry out Benedict's test on a sample to confirm a negative result.

- Hydrolyse another sample by heating it with dilute acid (e.g. HCl) or by using the enzyme **sucrase** (invertase) at its optimum temperature.

- When cooled, add dilute **sodium hydroxide** solution (NaOH) to neutralise the acid.

- Add **Benedict's solution** and heat in a water bath for 5 minutes.

- A positive **brick red** colour indicates a **non-reducing sugar** (sucrose) was present in the sample.

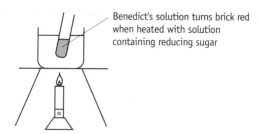

Benedict's solution turns brick red when heated with solution containing reducing sugar

Benedict's test

✓ *Quick check 3*

? Quick check questions

1 Name the bond used to join two glucose molecules together.

2 Explain how a disaccharide is formed from two monosaccharides.

3 Describe how you would test a sample of apple juice for the presence of reducing sugar.

B HB Carbohydrates – polysaccharides

These polysaccharides are large polymers made from monosaccharide monomers. The polysaccharides cellulose, starch and glycogen are polymers of glucose, and are formed by joining many glucose monomers (molecules) by condensation reactions.

- Polysaccharides differ in the number and arrangement of the glucose molecules they contain.

- They function as storage or structural molecules.

- They are not sweet to taste and are relatively insoluble in water.

> Polysaccharides are **non-reducing** – giving a negative result in Benedict's test.

Cellulose

Cellulose is found in the cell wall of plants, which provides rigidity and shape to the cell.

- Cellulose has thousands of β glucose molecules joined together by glycosidic bonds produced by condensation reactions to form a **long, straight chain**.

- **Hydroxyl** (OH) groups on each chain form weak **hydrogen bonds** with hydroxyl groups of other chains, producing a **microfibril**.

- The long, straight chains allow cellulose molecules to lie parallel to each other and form many hydrogen bonds, which hold the molecules firmly together – making microfibrils very strong.

> Cellulose is hydrolysed by the enzyme **cellulase**.

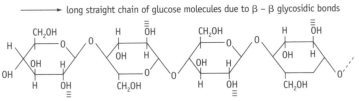

long straight chain of glucose molecules due to β – β glycosidic bonds

OH groups form hydrogen bonds to other cellulose molecules – which are all very long and straight ⟶ many hydrogen bonds

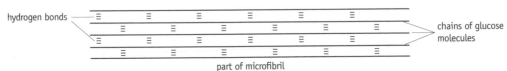

Structure of cellulose

- Microfibrils are grouped into larger bundles known as **macrofibrils**.

- Macrofibrils in one layer are orientated in the same direction.

- Macrofibrils in successive layers are orientated in a different direction.

- Macrofibrils of these different layers are interwoven and embedded in a matrix – making the cell wall **rigid**.

- The cellulose cell wall is usually **fully permeable** due to small channels between the different layers of macrofibrils.

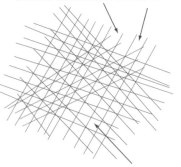

layers of macrofibrils orientated in different directions and embedded in a matrix

gaps provide full permeability

Arrangement of macrofibrils

✓ *Quick check 1*

Starch

This is the storage carbohydrate found in plants, consisting of long, **branched** chains of α-glucose molecules. Starch is stored in starch grains (**amyloplasts**) in the cytoplasm. Starch is ideally suited to its function as a storage compound because:

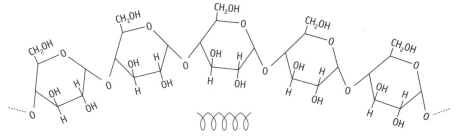

helical shape of molecule due to $\alpha - \alpha$ glycosidic bonds

Structure of starch

- it is **insoluble** and therefore does not affect the **water potential** of the cell and water movement due to osmosis (it is osmotically inactive);

- the molecule has a **helical** shape, forming a **compact store**;

- it contains a large number of **glucose molecules** providing a plentiful supply of **respiratory substrate**;

- it is too large to cross the cell membrane and so stays where it is formed.

> Foods obtained from plants usually test positive for starch. Starch is not stored in animal/human cells.

Example: starch is hydrolysed by the enzyme amylase to produce the disaccharide maltose.

Starch	+	water	$\xrightarrow[\text{hydrolysis}]{\text{amylase}}$	maltose
polysaccharide				disaccharide

✓ *Quick check 2*

Starch can be detected in a sample by using the iodine test. This involves:

- adding two or three drops of potassium iodide solution;

- if starch is present a blue/black colour is produced;

- if no starch is present the iodide solution remains orange/yellow.

Glycogen

This is the storage carbohydrate found in the cytoplasm of animal cells – sometimes referred to as 'animal starch' due to similarities in structure and function between the two molecules. Glycogen:

- is a polymer of α-glucose, similar to starch but with more branches;

- has the same storage advantages as starch;

- is stored in large amounts in liver and muscle tissues.

✓ *Quick check 3*

? Quick check questions

1 Explain how the structure of the cellulose molecule leads to the rigidity of a plant cell wall.

2 Give two features of a starch molecule that are related to its function as a storage compound.

3 Name one tissue that contains a large amount of glycogen.

B
HB

Proteins

Proteins contain carbon, hydrogen, oxygen, nitrogen and sometimes sulphur.
Amino acids are the monomers from which proteins (polymers) are formed.

- There are 20 different commonly occurring amino acids in living organisms.
- All amino acids have an amino group and a carboxylic acid group but differ in the structure of their R groups.
- Amino acids are joined together by **condensation** reactions – **peptide bonds** are formed.
- Two amino acids are joined together to form a **dipeptide**.
- Many amino are acids joined together to form a **polypeptide**.
- A protein may consist of one or more polypeptides.
- Proteins can be hydrolysed by heating with acid or by enzymes (proteases).

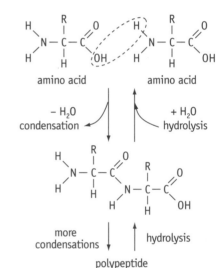

Generalised structure of an amino acid

✓ *Quick check 1, 2*

Biuret test for proteins

Protein can be detected in a sample by these steps:

- It is added to a test tube containing 2 cm^3 of dilute **sodium hydroxide** solution.
- Dilute **copper sulphate** solution is then added drop by drop.
- A **purple**, **lilac** or **mauve** colour indicates **protein** is present.
- If the solution remains blue, no protein is present.

Structure of proteins

Proteins vary in the number, type and arrangement of amino acids they contain. This produces a vast number of different protein molecules.

- **Primary structure** is the sequence of amino acids in the polypeptide chain of a particular protein.
- Specific interactions between different parts of the chain produce the final **3-dimensional** shape of the protein.
- **Secondary structure** is the folding or coiling of the polypeptide due to hydrogen bonding between amino acids in the chain.
- Secondary structures include the α-**helix** and the β-**pleated** sheet.
- **Tertiary structure** is the 3-dimensional shape of the protein due to folding and coiling of the secondary structure caused by specific **hydrogen bonds**, **ionic bonds** and **disulphide bridges** between parts of the chain.

Formation and hydrolysis of proteins

✓ *Quick check 3*

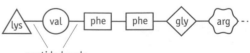

peptide bonds

Primary structure of a protein

> The tertiary structure or 3-dimensional shape of a protein is almost always the key to any property of the protein and any question!

Proteins can be classified according to their structure.

- A **fibrous protein**'s primary structure produces very large (insoluble) and long polypeptide chains with a simple tertiary structure.

- These wrap around each other to form strong, insoluble fibres, e.g. collagen in connective tissue.

- A **globular protein**'s primary structure produces a highly folded polypeptide chain with a very complex and specific tertiary structure.

- This allows them to recognise and bind specifically to other molecules (or ions), e.g. enzymes, antibodies and some hormones.

- Their amino acid composition, size and folding make them soluble and easy to transport.

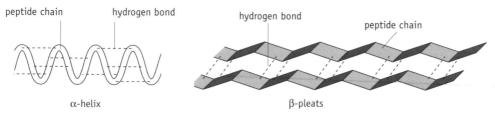

Secondary structures of proteins

Denaturation is a change in the tertiary structure of a protein – making it no longer functional.

- It is caused by temperatures above the optimum or extreme changes in pH.

- These conditions break hydrogen and ionic bonds.

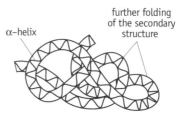

Tertiary structure of a protein

✓ *Quick check 4*

Quick check questions

1 Give one element present in all protein molecules which is not present in carbohydrates.

2 Give the generalised structural formula of an amino acid.

3 Explain what is meant by the primary structure of a protein.

4 Explain what happens to protein molecules when heated to high temperatures.

B

HB

Lipids

Lipids contain carbon, hydrogen and oxygen. They have a lower proportion of oxygen and a higher proportion of hydrogen compared to carbohydrates. They are used in respiration and as structural and storage molecules.

- Lipids are insoluble in water and thus do not affect the **water potential** of cells.
- They yield twice as much energy per gram as carbohydrates.
- They consist of fats (solids) and oils (liquids) at room temperature.

$$
\begin{array}{c}
\text{H} \\
|\\
\text{H} - \text{C} - \text{OH} \\
|\\
\text{H} - \text{C} - \text{OH} \\
|\\
\text{H} - \text{C} - \text{OH} \\
|\\
\text{H}
\end{array}
$$

Structure of a glycerol molecule

Triglycerides are one type of lipid formed by joining three **fatty acids** to one **glycerol** molecule by **condensation reactions**.

- The general formula of a fatty acid is **R–COOH**.
- All fatty acids have COOH, which is a carboxylic acid group.
- Different fatty acids have different R groups – representing long hydrocarbon chains which differ in the number of carbon atoms they contain and whether they are **unsaturated** (contain double bonds) or **saturated** (no double bonds).
- The three fatty acids in a lipid may be the same (simple lipid) or different (mixed lipid).

Formation and hydrolysis of a lipid molecule

✓ *Quick check 1*

Emulsion test for lipids

Lipid can be detected in a sample using the emulsion test, as follows:

- A small amount of the sample is placed into a test tube with 2 cm^3 of ethanol.
- The mixture is shaken so that the fat dissolves, then added to water in another test tube and shaken.
- A white or cloudy emulsion of fat droplets indicates fat is present.

✓ *Quick check 2*

Phospholipids

Phospholipids are lipids containing a phosphate group.

- A phospholipid molecule consists of one glycerol, two fatty acids and a phosphate group joined by condensation reactions.
- The phospholipid molecule has a polar **hydrophilic** head (attracts water) containing the phosphate group, and a non-polar **hydrophobic** tail (repels water) consisting of the long fatty acid chains.
- In cell membrane, phospholipids form a **bilayer** (two layers) as the hydrophobic tails are attracted to each other and away from water inside and outside the cell.

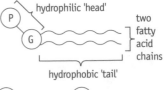

hydrophilic 'head'

two fatty acid chains

hydrophobic 'tail'

P phosphate   G glycerol

A phospholipid

✓ *Quick check 3*

? ## Quick check questions

1 What is a saturated fatty acid?

2 What result indicates the presence of lipid in the emulsion test?

3 How does a phospholipid molecule differ from a triglyceride?

Chromatography

Chromatography is used to separate and identify substances in a mixed solution.

Paper chromatography

- Solvent moves up the paper by capillary action, with substances in solution.
- The more soluble a substance is, the further it moves with the solvent.
- Before the solvent reaches the top, the **solvent front** is marked.
- The paper is dried and locating agents used to stain colourless substances.
- Each substance shows as a spot, whose distance moved can be measured.

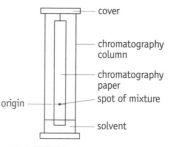

Paper chromatography

Identification of substances

A substance can be identified by its R_f value, which is constant in a specific solvent. The position of a spot can also be compared to the position of spots of known substances (markers) on the same paper/chromatogram.

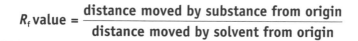

$$R_f \text{ value} = \frac{\text{distance moved by substance from origin}}{\text{distance moved by solvent from origin}}$$

▶ Insoluble compounds such as starch remain at the origin during chromatography.

✓ **Quick check 1, 2**

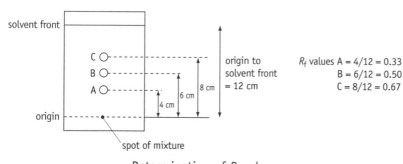

R_f values A = 4/12 = 0.33
B = 6/12 = 0.50
C = 8/12 = 0.67

Determination of R_f values

Two-way chromatography

Some substances have similar R_f values in a particular solvent and their spots overlap. The paper can be turned through 90° and put into a different solvent. The substances will travel different distances in the second solvent and give separate spots.

▶ Paper chromatography can be used in a laboratory to separate and identify: the different pigments of chlorophyll; the amino acids present in a hydrolysed protein; different monosaccharides and disaccharides.

✓ **Quick check 3**

? Quick check questions

1 Give the formula used to calculate R_f values.
2 Explain why locating agents are sometimes required.
3 Explain how you could use chromatography to ensure the separation and identification of two similar compounds.

B HB Enzymes

Enzymes are **globular proteins**. As **biological catalysts**, enzymes:

- allow biochemical reactions to happen at the temperature found in an organism;
- are unaffected by reactions and can be used many times;
- have a specific tertiary structure and react with a specific **substrate** to produce a specific **product.**

Vital chemical reactions in living organisms need specific enzymes to proceed, so specific enzymes regulate biological processes.

> The key to understanding enzymes is to remember they are proteins with a very specific 3-D shape. This allows them to 'recognise' other substances that have a shape that fits specifically against some part of the enzyme.

Lowering the activation energy

Lowering the energy needed to start a reaction means that:

- chemical reactions in cells occur within an acceptable **temperature range**;
- the overall **rate of reaction** is increased.

Enzyme specificity

Enzymes are **highly specific**. Some act on a single substrate, others on particular chemical bonds.

- Enzyme specificity is due to the **tertiary structure** of an enzyme.
- This determines the shape (configuration) of the **active site.**

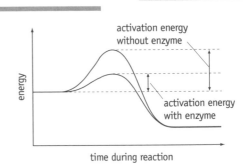

Lowering the activation energy

Lock and key hypothesis

The enzyme has an **active site** where a substrate with a complementary shape binds, forming an **enzyme–substrate complex**. The reaction takes place and the product is released, whilst the enzyme remains unchanged.

✓ **Quick check 1**

Lock and key hypothesis

Induced fit hypothesis

This is a modified model of the lock and key hypothesis. A substrate induces a change in the shape of an enzyme's active site so that they can fit together.

Effect of temperature

- An increase in temperature gives molecules more kinetic energy, resulting in **more collisions** between reactant molecules.
- The rate of reaction increases up to a maximum at the **optimum temperature**.
- Above the optimum, hydrogen and ionic bonds in the enzyme break and the enzyme is **denatured** – its **tertiary structure** has changed.
- The rate of reaction decreases as the substrate cannot attach to the active site.
- Denaturation of proteins at temperatures above 50°C is usually permanent.

> Read questions carefully – not all enzymes have an optimum temperature of 37°C and an optimum pH of 7!

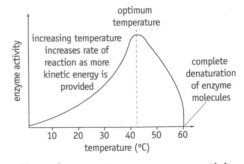

Effect of temperature on enzyme activity

Effect of pH

Enzymes possess an **optimum pH** at which the rate of reaction is at a maximum.

- Most enzymes work in a narrow pH range.
- pH changes alter ionic charges of acidic and basic groups.
- The tertiary structure and active site are altered, so substrate cannot bind.
- Extremes of pH can cause **denaturation** (e.g. due to acid hydrolysis of the enzyme).

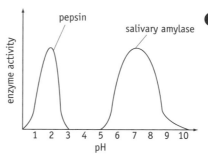

Effect of pH on rate of reaction

> Make sure you know that pH 7 is neutral, pH 1 is very acidic and pH 14 is very alkaline.

Effect of substrate concentration

Increasing the substrate concentration will increase the rate of reaction at first.

- The rate of reaction increases as collisions between substrate and enzyme molecules are more likely.
- The rate levels out as active sites of all the enzyme molecules are taken up by substrate molecules.
- The rate is then limited by the time required for the enzyme–substrate complex to form and release the product.
- Adding more enzyme will increase the rate.

Effect of substrate concentration on rate of reaction

✓ *Quick check 2, 3*

Enzyme inhibitors

Enzyme inhibitors slow down the rate of reaction.

A **competitive inhibitor**:

- has a **similar shape** to the substrate and **competes** for the active site;
- reduces the rate of a reaction, because the inhibitor occupies the active site.

This inhibition can be overcome by adding more substrate – increasing the chance of substrate occupying the active site.

A **non-competitive inhibitor**:

- does not have a similar shape to the substrate;
- binds at a site other than the active site, **altering the shape** of the enzyme and its active site;
- results in the substrate being unable to attach, or it binds but no product is formed.

The addition of more substrate will **not** reduce this inhibition.

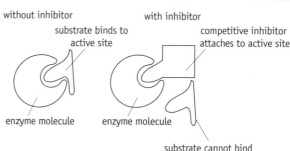

Competitive inhibition

Non-competitive inhibition

✓ *Quick check 4*

? Quick check questions

1. Explain why lipases can hydrolyse lipids but not carbohydrates.
2. Give three factors affecting the rate of an enzyme-catalysed reaction.
3. Explain in terms of molecular shape the effect of high temperature on enzyme activity.
4. Suggest a simple method by which you could determine whether a particular reaction is being inhibited by a competitive or non-competitive inhibitor.

B HB Tissues

In multicellular eukaryotic organisms, cells specialise or **differentiate** as they develop. They become adapted to carry out a particular function and this affects their shape and the number and type of organelles they contain.

A **tissue** consists of cells with the same adaptations and function, working together. An **organ** is a number of tissues working together to carry out a particular function.

Epithelial tissue

An epithelial tissue lines many organs and is especially important at exchange surfaces, where substances enter or leave an organ or organism. The **lung** is the organ where gaseous exchange takes place in microscopic air sacs called **alveoli.**

The features of the **alveolar epithelium** as a surface for gas exchange are:

- the **flattened shape** of individual cells and their large number – producing a **large surface area**;

- the presence of a **single layer** of **thin** cells – providing a **short diffusion pathway**.

Blood

Blood is a liquid tissue consisting of a number of different cell types, suspended in a liquid **plasma**.

White blood cells are part of the body's immune system – defending against invading microorganisms and foreign cells (including cancer cells).

- Each type of white cell has its own function and form.

- Examples include **lymphocytes**, **monocytes** and **granulocytes**.

✓ **Quick check 1**

► At exchange surfaces, substances have to cross cells and their partially/ selectively permeable plasma membranes in order to enter or leave an organism.

✓ **Quick check 2**

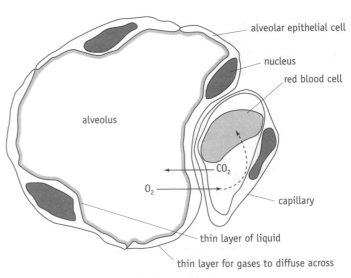

alveolar epithelial cell
nucleus
red blood cell
alveolus
CO_2
O_2
capillary
thin layer of liquid
thin layer for gases to diffuse across

Alveolar epithelium

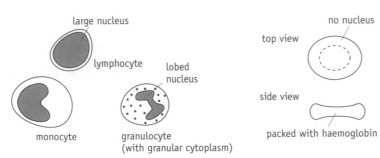

large nucleus
lymphocyte
lobed nucleus
monocyte
granulocyte (with granular cytoplasm)

no nucleus
top view
side view
packed with haemoglobin

White blood cells   Red blood cell

Red blood cells carry oxygen around the body. They contain **haemoglobin**, a red, iron-containing protein which binds reversibly to four oxygen molecules. Their adaptations include:

- **no nucleus** (when mature) – making more room for haemoglobin;
- flattened **bi-concave disc** shape, giving **a larger surface area to volume ratio**;
- large surface area for rapid loading and unloading of oxygen by diffusion;
- shorter diffusion pathways for oxygen.

✓ *Quick check 3*

Surface area to volume ratio

The relationship between the surface area of a cell/organism and its volume is important to the exchange of substances with the environment. Efficient exchange by diffusion needs:

- a large surface area relative to the volume of the cell/organism;
- thin exchange surfaces;
- short diffusion pathways;
- steep concentration gradients.

▶ It doesn't mean much just to write about a cell or organism having a large surface area – it has to be relative to the volume of the cell or organism.

Cells have a large surface area to volume ratio and short diffusion pathways to all parts of the cell. This is increased in red blood cells by flattening of the cell.

As the size of an organism increases, its surface area to volume ratio decreases. In large organisms diffusion pathways to tissues are too long to satisfy gas exchange requirements. Adaptations have evolved which maintain the adequate exchange of substances (e.g. nutrients, waste products, heat). These include:

surface area to volume ratio = 1.5 : 1
surface area to volume ratio = 2.5 : 1

4 cm

4 cm

4 cm

volume = 4 x 4 x 4 = 64 cm^3
surface area = (4 x 4) x 6 = 96 cm^2

8 cm

1 cm

8 cm

volume = 8 x 8 x 1 = 64 cm^3
surface area = (8 x 8) x 2 + (8 x 1) x 4 = 160 cm^2

Flattening increases surface area to volume ratio

- changes in body shape which increase the surface area for exchange – e.g. flattening of the body in flatworms, and large ears for elephants to provide a large surface area for losing heat.
- development of an internal respiratory system which has a large surface area relative to the volume of the organism.

✓ *Quick check 4*

? *Quick check questions*

1 Explain what is meant by: (**a**) a tissue; (**b**) an organ.

2 Explain how the alveolar epithelium is adapted for gaseous exchange.

3 Explain how the structure of a red blood cell is adapted for its transport function.

4 Flatworms are quite large organisms. Some are many centimetres in length and one or two centimetres in width. They are called flatworms because in cross-section their bodies are only a few millimetres deep. Suggest why flatworms do not need to have special respiratory systems for gaseous exchange.

B
HB
Blood vessels

Blood vessels are **organs** – structures made of different tissues. Blood flows rapidly under high pressure in **arteries**. These branch into smaller arteries, **arterioles** and then **capillary beds**. Exchange of substances with body tissues takes place in **capillaries**, where resistance to the flow of blood causes blood pressure and rate of flow to fall. Capillaries merge into **venules**, then **veins**, which carry blood slowly at low pressure towards the heart.

Arteries

Each heartbeat sends a surge of blood under **high pressure** out along arteries, pushing the blood forwards. The thick artery wall resists pressure and recoils.

- **Connective tissue** strengthens the outer layer.
- **Elastin fibres** in the **smooth muscle** layer have elastic properties.
- The lining layer is highly folded, to allow for expansion of the artery with each surge of blood.
- A layer of **endothelial tissue** one cell thick covers the inside of all blood vessels.

fibrous layer – to prevent splitting of artery wall due to high blood pressure

smooth muscle layer – thick muscle layer with elastin fibres, that resist expansion of artery wall and make it recoil to squeeze blood and maintain high pressure

lining layer – connective tissue and elastin fibres, with a lining layer of endothelial cells in contact with the blood

small lumen, carrying blood at high speed and pressure

Cross-section through an artery

Arterioles

Arterioles do not have to stand the very high pressure found in main arteries.

- The wall of an arteriole has an endothelial layer and smooth muscle tissue.
- Smooth muscle can contract or relax, decreasing or increasing blood flow into capillaries.
- Smooth muscle is under nervous control (sympathetic nervous system).

> Make sure you can label the layers of the artery and vein.

Veins

Veins carry blood under **low pressure**, so although their walls have the same layers and tissues as arteries, they do not have to be as thick.

- The lumen is very large, so even at low speed and pressure, blood flows back to the heart at the same rate as it leaves along the arteries.

- Pressure in veins is too low to lift blood from the lower body to the heart.

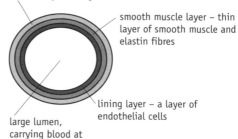

Cross-section through a vein

✓ Quick check 1, 2, 3

- Contracting muscles in the legs and body press on veins, squeezing blood along.

- Large veins have **semilunar valves** at intervals, to make sure that the blood travels in one direction.

- In A, blood is pushed up the vein and opens the semilunar valves (B).

- In C, blood flows back down the vein causing the valves to close.

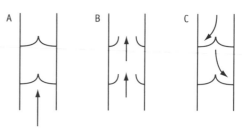

Valves in veins

? **Quick check questions**

1 Use information in the diagrams on the facing page to calculate how many times thicker the wall of the artery is compared with the vein. Assume the diagrams are to the same scale.

2 Explain how arteries and veins are adapted to carry blood at the same rate.

3 Suggest why we can feel a pulse where arteries run near the surface of the body.

> **◐** Make sure you know the function of each type of vessel and whether it carries blood towards or away from the heart.

B HB Circulation

Larger organisms like mammals have a small surface area to volume ratio. Specialised exchange systems are needed to get oxygen, water and nutrients from the environment and excrete wastes like carbon dioxide and urea. Exchange surfaces have large areas, which are thin and moist for rapid diffusion of substances in solution.

- In larger organisms, **exchange systems** work with **transport systems**.
- Transport systems move substances to and from exchange surfaces.
- This prevents build-up of substances at the exchange surfaces and **maintains concentration gradients**.
- **Mass flow systems** involve the movement of volumes of water (or gas) carrying substances **over large distances** through a transport system – the **bulk movement** of substances. For example, the **blood system** uses blood plasma (which is mainly water) to carry glucose, amino acids and carbon dioxide (as carbonate ions) in solution and blood cells in suspension.

✓ *Quick check 1*

General pattern of blood circulation

In mammals the blood flows in closed vessels: arteries, arterioles, capillaries and veins. It is moved through the vessels by the pumping action of the heart.

The pattern of blood circulation in humans is shown in the diagram.

✓ *Quick check 2*

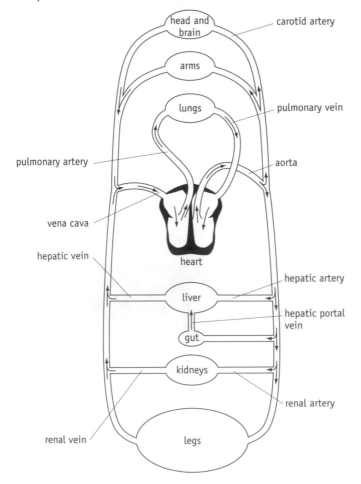

Pattern of blood circulation

▶ You need to know the names of only those blood vessels labelled in this diagram.

Capillaries and tissue fluid

Capillaries are where substances are exchanged between the blood and body tissues.

- Capillary walls are **one endothelial cell thick** – giving a very short pathway for the exchange of substances.
- There are **very large numbers** of capillaries – giving a large surface area for exchange.
- No cell in tissues is very far from a capillary – giving short diffusion pathways.

> ▶ Think of the capillaries as one, very large exchange surface – with all the characteristics of this kind of surface.

Tissue fluid and metabolic exchanges

Tissue fluid surrounds cells in tissues and is formed from substances which leave the blood plasma in capillaries.

Compared to plasma, tissue fluid has: a lower concentration of oxygen and higher concentration of carbon dioxide; no blood cells, platelets or large plasma proteins (they are too large to cross the endothelial cells of the capillary).

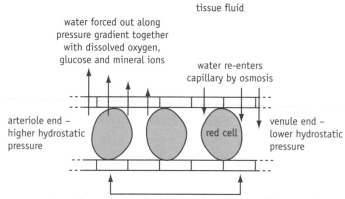

Formation of tissue fluid

- Substances involved in **metabolism** (the chemical reactions in the body), such as oxygen, glucose and mineral ions, diffuse into the tissue fluid along concentration gradients – maintained by cells constantly using them.
- Carbon dioxide and urea diffuse into the plasma along concentration gradients.
- Not all of the water leaving the capillary is reabsorbed.
- **Lymph** forms from surplus tissue fluid draining into lymph capillaries of the **lymphatic system**.
- These lymph capillaries are dead-end vessels which merge to form larger lymph vessels, which return lymph to the blood plasma.
- Lymph is similar to tissue fluid but can contain fats, more protein and white blood cells.

✓ *Quick check 3*

? *Quick check questions*

1 Explain why we need a blood system.
2 Cells in the kidney need a constant supply of glucose to maintain a high rate of respiration. Describe the route followed by a molecule of glucose from the gut to the kidney.
3 Explain how water is reabsorbed into a capillary from tissue fluid.

B
HB

Lung function

Lungs are the gas exchange system in humans. Lungs consist of:

- the **trachea**, which divides into two **bronchi** – all supported by rings of cartilage to prevent collapse;

- **bronchioles** – smaller tubes branching repeatedly from the bronchi;

- **alveoli** at the ends of the bronchioles – providing a large surface area where gaseous exchange occurs.

Exchange of respiratory gases in the lungs

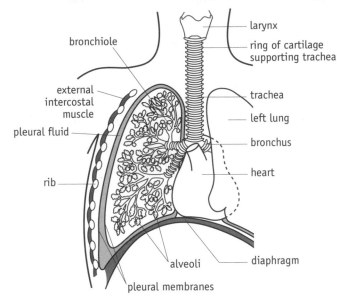

Structure of human lungs

Fick's law

$$\text{Diffusion rate} \propto \frac{\text{surface area} \times \text{difference in concentration}}{\text{thickness of exchange surface}}$$

High diffusion rates for oxygen and carbon dioxide happen when the exchange surface:

- has a large surface area;
- has a large difference in concentrations of the gases across it;
- is very thin.

The **alveolar epithelium** has:

- large numbers of flattened cells – producing a large surface area;
- a single layer of thin cells – providing a short diffusion pathway.

Ventilation (breathing) keeps the alveolar concentrations of oxygen high and carbon dioxide low.

- Oxygen diffuses along its **diffusion gradient** across the alveolar epithelium and the capillary endothelium into the blood plasma and then into red blood cells – where it binds to haemoglobin.
- Carbon dioxide diffuses in the opposite direction.
- Blood circulating in the capillaries removes absorbed oxygen and brings carbon dioxide – **maintaining high/steep diffusion gradients.**

Ventilation

Organisms with internal gas exchange surfaces need to get gases to and from these surfaces – ventilation systems are needed.

In mammals, **inspiration** (inhalation, breathing in) is an active process.

- Intercostal muscles contract – moving the ribcage upwards and outwards.
- Diaphragm muscles contract causing the diaphragm to flatten.

> **Learn by heart** the equation for Fick's law and the common features of all exchange surfaces.

A group of alveoli

> **branch of pulmonary vein**
> **branch of pulmonary artery**
> **capillary network**
> **alveoli**

✓ *Quick check 1*

> Never say that air is sucked into the lungs! It is forced into the air passages – the trachea, bronchi and bronchioles – by higher external atmospheric pressure.

- These actions increase the volume of the thorax.
- The pressure inside the thorax decreases below atmospheric pressure and air enters the lungs along a pressure gradient.

Expiration (breathing out) is mainly a passive process.

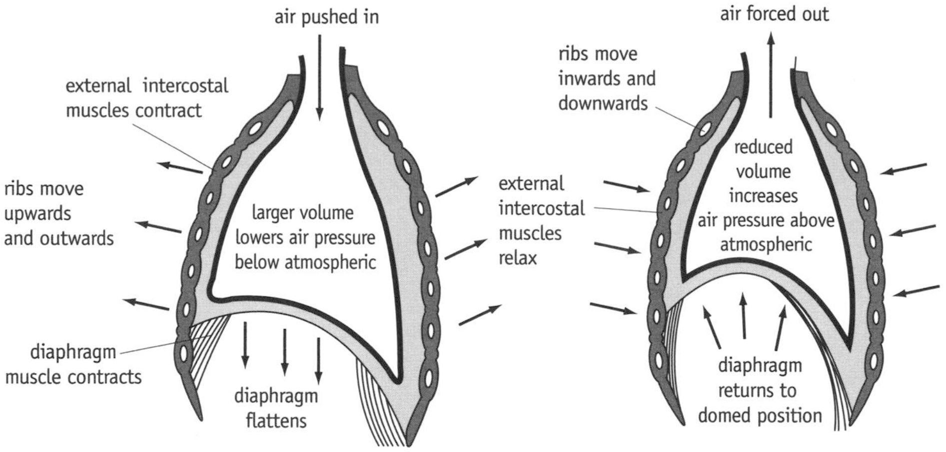

air pushed in

external intercostal muscles contract

ribs move upwards and outwards

larger volume lowers air pressure below atmospheric

diaphragm muscle contracts

diaphragm flattens

air forced out

ribs move inwards and downwards

reduced volume increases air pressure above atmospheric

external intercostal muscles relax

diaphragm returns to domed position

Ventilation in mammals

Composition of inhaled and exhaled air

✓ *Quick check 2*

- The body does not use nitrogen – its percentage changes because of changes in other gases.
- Oxygen is absorbed for respiration and carbon dioxide excreted.

Gas	% in inhaled air	% in exhaled air
Oxygen	20.9	15.3
Carbon dioxide	0.03	3.7
Water vapour	0 in dry air	6.2
Nitrogen	79.0	74.8

Control of ventilation

Breathing is controlled by the **medulla** in the brain, containing a respiratory centre, divided into **inspiratory/inhalation** and **expiratory/exhalation** centres.

- The inspiratory centre causes intercostal and diaphragm muscles (effectors) to contract, it inhibits the expiratory centre and we breathe in.
- As the lungs inflate, **stretch receptors** send **nerve impulses** to the medulla, **inhibiting** the inspiratory centre (Hering–Breuer reflex).
- The expiratory centre is no longer inhibited and we exhale.
- As the lungs deflate, stretch receptors become inactive, the inspiratory centre becomes active and the expiratory centre is inhibited.
- These events produce a basic breathing rhythm.

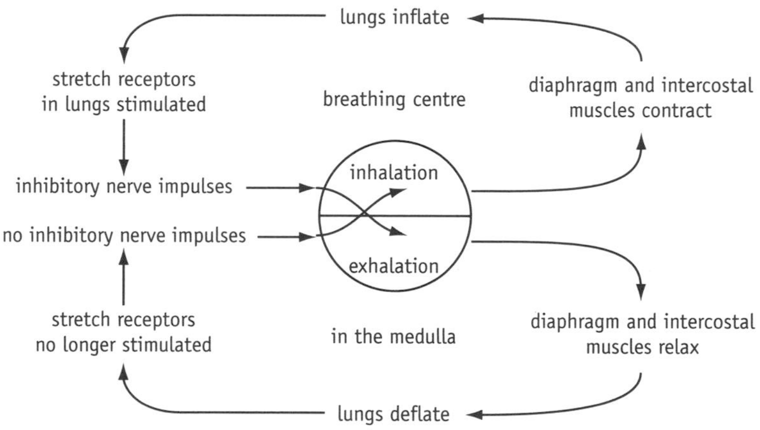

lungs inflate

stretch receptors in lungs stimulated

breathing centre

diaphragm and intercostal muscles contract

inhibitory nerve impulses

inhalation

no inhibitory nerve impulses

exhalation

stretch receptors no longer stimulated

in the medulla

diaphragm and intercostal muscles relax

lungs deflate

Control of ventilation

You can use the flow chart to help you remember facts but it isn't an 'explanation' on its own. Try to explain in writing what the chart shows.

✓ *Quick check 3*

? *Quick check questions*

1 Give two ways in which a respiratory surface may be adapted for gaseous exchange.

2 Describe how inhalation takes place in mammals.

3 Explain how a basic breathing rhythm is maintained.

Heart structure and function

B
HB

The blood vascular system circulates blood around the body in blood vessels. Blood is pushed through the vessels by the pumping action of the heart.

- **Atria** – receive low-pressure blood returning to the heart in **veins**.

- **Ventricles** – thick muscle to pump blood at high pressure into **arteries**.

- The left ventricle has a thicker wall, because it pumps blood to most of the body.

- Blood flows in **one direction** through the heart and blood vessels.

- **Atrioventricular valves** open when ventricles relax and close as they contract.

- **Semilunar valves** open as the ventricles contract.

- They close as the ventricles relax, preventing backflow into the ventricles.

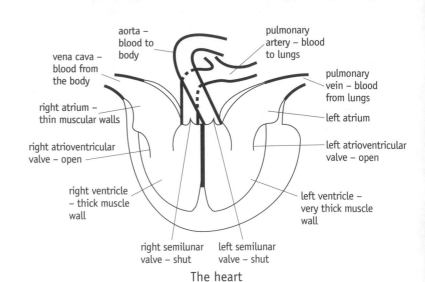

The heart

Myogenic stimulation of the heart

The heart is **myogenic** – it beats regularly without input from the nervous system.

Sinoatrial node, atrioventricular node and the heartbeat

- The **sinoatrial node** (SA node) is a patch of modified muscle cells in the wall of the **right atrium** – it produces regular bursts of electrical impulses.

- These impulses spread rapidly as waves of electrical activity through the walls of the right and left atria, making them contract together.

- The impulses reach the **atrioventricular node** (AV node).

- There is a **delay** of 0.15 seconds before the AV node reacts, so that the ventricles contract after the atria.

- Impulses from the AV node travel rapidly through the **bundle of His** and into branches to all parts of the ventricles.

- The ventricles contract, starting at the bottom, to push the blood up and out into the arteries.

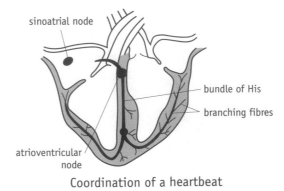

Coordination of a heartbeat

> Make sure you can label the parts of the heart.

> ✓ **Quick check 1**

> Make sure you can explain how the order of beating of the atria and ventricles and the opening and closing of the valves make blood flow in one direction through the heart.

> ✓ **Quick check 2**

> There are **two** nodes involved in the heartbeat.

The cardiac cycle

This is the sequence of contraction and relaxation of the heart chambers, and opening and closing of valves, during one heartbeat.

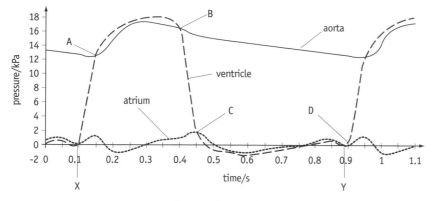

The cardiac cycle

▶ The sequence of events involving the nodes and the bundle of His during one contraction cause blood to flow in **one** direction through the heart.

The graph above shows pressure changes in the left atrium, left ventricle and aorta during one cardiac cycle. X and Y mark one cardiac cycle; the graph starts to repeat itself after Y.

- Time A – the left ventricle is contracting and pressure inside increases to **above** that in the aorta.
- The semilunar valve opens – blood flows into the aorta and the volume of the ventricle decreases.
- Time B – the ventricle is relaxing and elasticity makes its volume increase, reducing the pressure inside to **below** that in the aorta.
- The semilunar valve closes, preventing backflow of blood into the ventricle.
- Time C – pressure inside the ventricle is falling below that in the atrium and the atrium is contracting, producing a small pressure.
- The atrioventricular valve opens, allowing blood to flow into the ventricle, and the volume of the atrium decreases.
- Time D – the ventricle is contracting again and the pressure inside rises above that in the atrium, making the atrioventricular valve close and prevent backflow of blood.

▶ Make sure you understand how changes in the pressure of the blood open and close valves. You should be able to interpret graphs like the one shown on this page.

? Quick check questions

1 Explain how the structure of the heart causes blood to flow in one direction.
2 Explain why the time delay before electrical impulses from the sinoatrial node reach the atrioventicular node is important to the flow of blood from the atrium to the ventricle.
3 Use the data in the graph above to calculate the rate at which the heart is beating.
4 Explain why blood does not flow from the right ventricle into the right atrium.

B
HB

Effects of exercise

During exercise muscles use more energy. The flow of blood to muscles increases, bringing more oxygen and glucose for respiration and getting rid of more carbon dioxide and lactic acid.

The flow of blood to muscles depends on cardiac (heart) output and the distribution of blood flow.

Cardiac output and pulmonary ventilation

Cardiac output (cm^3 per min)	=	stroke volume (cm^3 per beat)	×	heart rate (beats per min)

- **Stroke volume** – the volume (cm^3) of blood pumped per beat by each ventricle.
- At rest, it is usually between 70 and 80 cm^3 per beat in adults.
- Resting heart rate is about 70 beats per minute.
- This gives a **cardiac output** of between 4900 and 5600 cm^3 per minute (4.9 to 5.6 litres per minute).

Pulmonary ventilation (breathing) affects the amount of oxygen entering the blood and carbon dioxide leaving.

Pulmonary ventilation (cm^3 per min)	=	tidal volume (cm^3)	×	breathing rate (per minute)

> ▸ Make sure you know the definitions of and units for all of the terms relating to cardiac output and pulmonary ventilation.

- **Tidal volume** – the volume of each breath when at rest.

✓ *Quick check 1*

Changes with exercise

During exercise, **cardiac output increases** – due to increases in heart rate and stroke volume.

- Heart rate is controlled by a **cardiac centre** in the **medulla** of the brain, divided into a cardioaccelerator centre and a cardioinhibitor centre.
- Contracting muscles pressing on veins force blood towards the heart, causing greater filling of the ventricles and making the heart beat faster, stronger and with a greater stroke volume.
- The main effect is an increase in heart rate.
- If blood pressure rises too far above normal, **pressure receptors** in the aortic and carotid sinuses send nerve impulses to the cardioinhibitor centre.
- This centre sends inhibitory nerve impulses to the cardioaccelerator centre and the SA node – preventing the heart beating too fast.

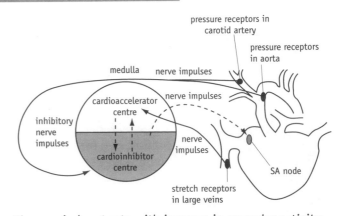

Changes in heart rate with increase in muscular activity

> ▸ Don't confuse how heartbeat is regulated with the control of breathing!

✓ *Quick check 2*

During exercise, **pulmonary ventilation increases –** due to increases in the depth and rate of breathing.

- Breathing is controlled by a **breathing/respiratory centre** in the **medulla** of the brain.
- Increased muscle activity produces more carbon dioxide from respiration.
- This is converted to carbonic acid in red blood cells, and lowers blood pH.
- **Chemoreceptors** stimulated by lower pH are found in the **aortic body**, **carotid body** (in the aorta and carotid artery) and the **medulla**.
- Nerve impulses travel to the medulla, leading to increases in the rate and depth of breathing and rate of excretion of carbon dioxide.
- After exercise, the rate of breathing remains high until the concentration of carbon dioxide in the blood falls to normal.

> Make sure you can use the terms used here to explain why breathing rate increases during exercise – simpler terms won't do!

✓ *Quick check 3*

Redistribution of blood flow with exercise

During exercise, flow of blood to the skeletal muscles **increases** because of:

- increased cardiac output;
- increased opening of blood vessels in the muscles (vasodilation);
- **redistribution of blood** away from some other parts of the body – especially the gut and skin.

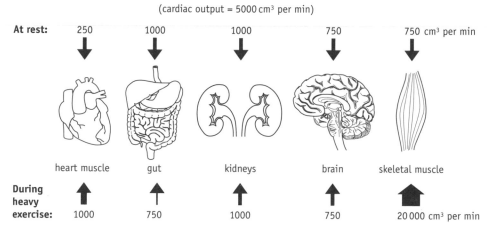

(cardiac output = 5000 cm³ per min)

At rest: 250 1000 1000 750 750 cm³ per min

heart muscle   gut   kidneys   brain   skeletal muscle

During heavy exercise: 1000 750 1000 750 20 000 cm³ per min

(cardiac output = 25 000 cm³ per min)

Redistribution of blood flow in response to exercise

Stability of some blood supplies

Some organs need a stable blood supply (to supply enough oxygen and glucose for respiration), to work efficiently whatever the body is doing.

During exercise, a **cut in blood supply** to:

- **the heart** – would starve heart muscle of oxygen and glucose, making it unable to pump more blood, and might cause a heart attack;
- **the brain** – would reduce ability to react to danger and might result in unconsciousness/death;
- **the kidneys** – would cause a build-up of toxins in the blood.

> Make sure that you can calculate percentage changes in blood supply to organs during exercise.

✓ *Quick check 4*

? Quick check questions

1 Explain what is meant by: **(a)** cardiac output; **(b)** pulmonary ventilation.

2 Explain how cardiac output changes during exercise.

3 Explain how and why pulmonary ventilation changes during exercise.

4 **a** Use the data in the diagram above to calculate the percentage changes in blood supply to the gut and skeletal muscles during exercise.

b Explain the reasons for these changes.

c After a large meal, we often feel sleepy. Using your knowledge of possible changes in blood supply to organs, suggest an explanation for this sleepiness.

Module 1: end-of-module questions

1 a Explain how the structure of each of the following carbohydrates is related to its function: (i) cellulose; (ii) starch. [4]

b Use diagrams to describe how fatty acids and glycerol combine to form a triglyceride. [3]

c Explain how the primary structure of a polypeptide leads to its tertiary structure. [4]

2 a Describe how you would test a solution for the presence of each of the following substances: (i) sucrose; (ii) protein; (iii) lipid. [3]

b You are given a solution containing a soluble protein. Suggest how you would find out which amino acids are present in this protein. [4]

3 a i Describe the differences you would expect to see between an animal cell and a plant cell when viewed with a light microscope. [3]

ii Name three organelles that you would expect to see in both types of cell with an electron microscope and describe the function of each. [6]

b When scientists first saw mitochondria with an electron microscope they had no idea what their function was. Explain how they were able to find the function of mitochondria in cells that contain many different types of organelle. [3]

4 a The diagram shows three plant cells and their water potentials.

i Use arrows to show the directions of net water movement between these cells. [1]

ii Explain why net movements of water would take place. [3]

b The chart shows concentrations of ions in the sap of a plant cell and in the solution surrounding the cell.

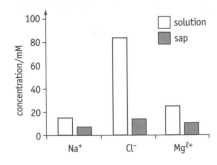

i Explain the concentration of chloride ions in the sap compared with the external solution. [2]

   ii Suggest why the different ions accumulate in the cell sap by different amounts. [2]

   iii The cell was placed into a new solution with the same concentrations of ions but half the concentration of dissolved oxygen. Suggest the effect that this would have on the results shown above. [3]

5 a Explain what is meant by each of the following:

   i diffusion; [2]

   ii facilitated diffusion; [3]

   iii active transport. [3]

b Explain why water can cross the cell membrane by diffusion but glucose needs a carrier protein. [5]

6 a Explain the importance of enzymes to the chemical reactions necessary to support life. [3]

b The graph shows the effects of increasing substrate concentration on the rate of action of an enzyme with and without an inhibitor.

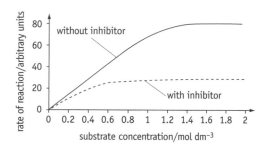

   i Explain the results without the inhibitor. [3]

   ii Describe and explain the effect of the inhibitor. [3]

7 The graph shows the effect of temperature on the activity of an enzyme.

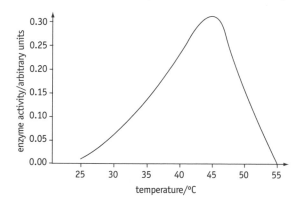

a Give the optimum temperature for this enzyme. [1]

b i Explain the activity of the enzyme at 25°C. [2]

   ii Explain the activity of the enzyme at 55°C. [3]

c Two samples of the enzyme were taken, both were stored at 5°C for several days but one was boiled before being stored. Both samples were then warmed to the optimum temperature for the enzyme. Substrate was added and the activity of the enzyme in each sample was measured.

B

HB

   i Explain any differences in the results you would expect with
     each sample. [2]

   ii Explain three factors you would need to keep constant to make
      accurate comparisons about the effects of temperature on the samples. [6]

8 a Explain how the contraction of the atria and ventricles is controlled
to make sure that both sides of the heart beat together and that
blood flows in the correct direction through the heart. [6]

b The heart beats with its own rhythm but can speed up during muscular
activity.

   i Explain why the heart needs to speed up. [3]

   ii Explain how the speeding up of the heart happens. [6]

9 a Explain how the structures of the heart make blood flow in one direction. [4]

b The graph shows changes in pressure in the left ventricle and aorta
during one cardiac cycle.

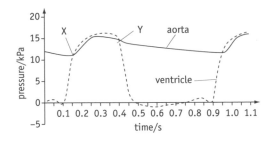

   i Explain what happens at points X and Y. [4]

   ii Calculate the heart rate in beats per minute. [2]

Module 2: Making Use of Biology

This module is broken down into seven topics: Cell division; The genetic code; Gene technology; Enzymes from microorganisms; Forensic examination of blood; Cultivated plants; Biotechnology and reproduction. **The first three topics (pages 36–47) are also required for pupils taking Module 3 (Human Biology).**

Cell division B HB

- Growth and asexual reproduction depend on mitosis, producing genetically identical cells.
- In sexual reproduction gametes fuse at fertilisation. Meiosis halves the number of chromosomes in cells. Fertilisation reverses this and restores the usual chromosome number.

The genetic code B HB

- DNA carries genetic information from one generation to the next.
- The structure of DNA allows it to carry coded genetic information and replicate itself. Specific base pairing is involved in replication of DNA and use of information by the cell.
- A gene is a length of DNA carrying information for making a specific protein (often an enzyme).

Gene technology B HB

- Genetic engineering allows us to remove genes from one organism and insert them into another (even if it is a different species) – making it genetically engineered, or modified.
- This recombinant DNA technology raises ethical problems.

Enzymes from microorganisms

- Enzymes can be isolated and purified from microorganisms.
- High sensitivity and specificity make them useful analytical reagents.
- Thermostable enzymes can be obtained for industrial processes. Immobilised enzymes are more stable and easily controlled than mobile ones.

Forensic examination of blood

- White blood cells respond to foreign antigens. Understanding of this immunological response allows us to determine someone's ABO blood group.
- DNA can be replicated using the polymerase chain reaction and then be used to produce a genetic fingerprint.

Cultivated plants

- Cereals are important in the human diet.
- Where a cereal grows depends on its adaptations to climate.
- Humans can change the abiotic environment to help plants.
- Attempts to help crops can have adverse environmental impacts.

Biotechnology and reproduction

- Understanding of hormonal control of reproduction allows us to manipulate and control reproduction in humans and other mammals.

Mitosis

Mitosis is the type of cell division that produces genetically identical cells. During mitosis DNA replicates in the parent cell, which divides to produce two new cells, each containing an exact copy of the DNA in the parent cell. The only source of genetic variation in the cells is via mutations.

- Mitosis increases cell numbers during growth, repair of tissues and asexual reproduction.

- During mitosis, the nuclear material becomes visible as structures called **chromosomes**.

- In a normal body cell (somatic cell) the chromosomes can be grouped into **homologous pairs** of chromosomes.

- **Diploid number (2n)** is the total number of chromosomes in a normal body cell. In humans this is 46, i.e. 23 homologous pairs.

- **Haploid number (n)** is the number of chromosomes in a **single set,** i.e. one member from each homologous pair. In humans this is 23, the number of chromosomes in a **gamete** (sperm or ovum).

- Mitosis produces cells with the same number of chromosomes as the parent cell so that a diploid parent cell will divide to produce two identical diploid cells.

> The diploid number of chromosomes is a characteristic of a particular species – and it isn't usually 46!

> Homologous chromosomes carry the same genes but may not carry the same alleles of the genes.

> ✓ *Quick check 1, 2*

> In rapidly dividing cells mitosis is completed within 24 hours.

Process of cell division

Biologists have divided the process of cell division into a number of stages: interphase, prophase, metaphase, anaphase and telophase.

Interphase

- This is when the cell is **not dividing,** but is carrying out its normal cellular functions.

- During interphase the cell prepares for nuclear division by:

  - **DNA replication**, doubling the genetic content of the cell

  - increasing its **ATP** content, as nuclear division is a very active process.

- Replication of cell organelles, e.g. mitochondria, occurs in the cytoplasm.

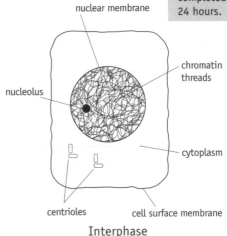

Interphase

Prophase

- The previously indistinct nuclear material is now visible as **chromosomes**.

- Due to DNA replication during interphase, each chromosome consists of two identical **sister chromatids** connected at the **centromere**.

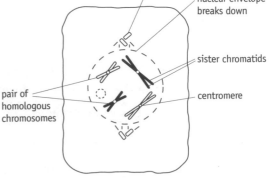

Prophase

- Each chromosome shortens and thickens, a process known as **condensation**.
- Condensation of chromosomes prevents tangling with other chromosomes.
- Centrioles (in animal cells) move to opposite poles (sides) of the cell.
- The nucleoli and **nuclear membrane** break down.

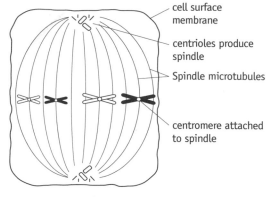

cell surface membrane

centrioles produce spindle

Spindle microtubules

centromere attached to spindle

Metaphase

Metaphase

- A **spindle** (of protein microtubules) forms across the cell.
- Each chromosome moves to the equator of the spindle and attaches to a spindle fibre by its **centromere**.
- The **sister chromatids** of each chromosome are orientated towards opposite poles of the cell.

Anaphase

- The centromere splits and the **sister chromatids separate**.
- Sister chromatids move to opposite poles of the **spindle**.
- Numerous mitochondria around the spindle provide energy for movement.

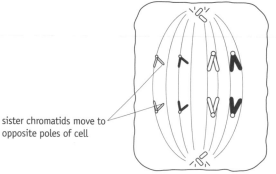

sister chromatids move to opposite poles of cell

Anaphase

Once the chromatids reach the poles of the spindle they are called chromosomes. They will replicate themselves during the next interphase to produce two sister chromatids – ready for the next cell division.

Telophase

- The chromatids are at opposite poles of the cell.
- A nuclear membrane forms around each set of chromatids.
- The two nuclei formed are **genetically identical** to each other and the original parent cell.
- Two new cells are formed as **cytoplasmic cleavage** occurs and a cell membrane forms between the cells.

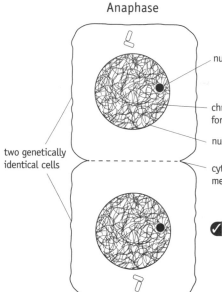

nucleolus forms

chromatids unwind to form chromatin

nuclear envelope forms

two genetically identical cells

cytoplasm divides and cell membrane forms

✓ *Quick check 3*

Telophase

? *Quick check questions*

1 Give two ways in which mitosis is important in living organisms.

2 Explain what diploid number means.

3 During which stage of mitosis does each of the following occur:
   (**a**) separation of chromatids; (**b**) condensation of chromosomes; (**c**) DNA replication?

B
HB

Mitosis and the cell cycle

The cell cycle is the period from a cell being formed by division to the point when that cell itself divides. Most types of cell never divide after they become specialised, but unspecialised cells often undergo cell division. The length of the cell cycle can vary considerably. Cells lining the gut have a very short cell cycle, as cell division continually occurs to replace cells lost as food passes through.

The cell cycle

- Consists of interphase, mitosis and cytokinesis.
- **Interphase** consists of three stages: G_1, S, and G_2.
- During G_1 (**first growth phase**) new proteins and organelles are made – cell growth occurs.
- In the **S phase** DNA replication occurs.
- During G_2 (**second growth phase**) more organelles and proteins required for mitosis are made.
- **Mitosis** then occurs, producing two identical nuclei.
- **Cytokinesis** is the division of the cytoplasm and the formation of two identical cells.

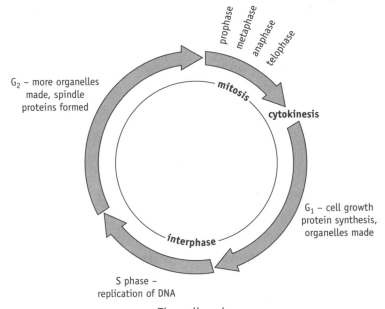

The cell cycle

✓ *Quick check 1, 2*

Most cells have a limited number of cell cycles but cancer cells can continually divide.

Observation of mitosis

The stages of mitosis can readily be observed in the zone of cell division in a root tip. The chromosomes are stained with a dye such as acetic (ethanoic) orcein.

- Cut the end 2 mm from a growing root tip (e.g. onion bulb).
- Place this in a glass dish containing a mixture of acetic (ethanoic) acid and alcohol (1:1) for 10 minutes.
- Wash with water and transfer to another dish containing HCl and warm for 5 minutes.
- HCl helps separate the cells.
- Wash and place tip in acetic (ethanoic) orcein on a slide.
- Put on a coverslip and gently squash the root tip using filter paper to produce a thin layer and to help separate the cells.
- Examine the slide for stages of mitosis.

Observing mitosis

✓ *Quick check 3*

? *Quick check questions*

1 What are the three main stages of the cell cycle?

2 What takes place during the S phase of the cell cycle?

3 Name a stain used to dye chromosomes.

Meiosis

During meiosis in diploid organisms such as humans, cells containing pairs of homologous chromosomes divide to produce **haploid gametes**, containing one chromosome from each homologous pair. The cells produced are genetically different from the parent cell and from each other.

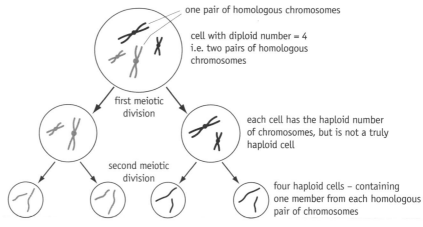

one pair of homologous chromosomes

cell with diploid number = 4 i.e. two pairs of homologous chromosomes

first meiotic division

each cell has the haploid number of chromosomes, but is not a truly haploid cell

second meiotic division

four haploid cells – containing one member from each homologous pair of chromosomes

Process of meiosis

In meiosis:

- the DNA in a cell replicates only once;

- but the cell then divides twice;

- the number of chromosomes is reduced from the diploid number ($2n$) to the haploid number (n);

- four new cells are formed which are **genetically different** from each other;

- the cells produced usually function as **gametes** (i.e. reproductive cells).

Importance of meiosis

- In diploid organisms, meiosis is important in sexual reproduction as it ensures that **haploid gametes** are produced.

- These fuse at **fertilisation** to form a **zygote** and the **diploid number** is restored.

- Meiosis ensures that each generation possesses a **constant number of chromosomes** (e.g. 46 chromosomes in humans).

- If meiosis did not occur and diploid gametes were produced, the number of chromosomes would double every generation after fertilisation.

- The process of meiosis produces genetic variation in gametes.

✓ *Quick check 1, 2*

Do not learn lots of detail about what happens in meiosis – it will not be asked about in the AS exam. These details are in the A2.

✓ *Quick check 3*

1 How many cells are produced when a single cell undergoes meiosis?

2 What is a gamete?

3 Explain how meiosis enables a constant number of chromosomes to be maintained from generation to generation.

B
HB

Structure of nucleic acids

DNA (deoxyribonucleic acid) and RNA (ribonucleic acid) are nucleic acids. They are **polymers** of **nucleotides**. DNA consists of two **polynucleotide** strands, whereas RNA consists of a single polynucleotide strand. Each nucleotide consists of three molecules joined by **condensation** reactions:

* a **five-carbon sugar** (pentose);
* a **phosphoric acid** molecule;
* a **nitrogen-containing organic base**.

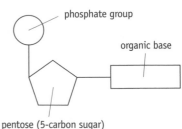

Structure of a nucleotide

DNA

Structure of DNA

The two polynucleotide strands are held together by **hydrogen bonding** – forming a **double helix**. In DNA, four bases are found in the nucleotides: **cytosine** and **thymine** (pyrimidines); **adenine** and **guanine** (purines).

* The sugar **deoxyribose** and **phosphate** form the backbone of the polynucleotide strands.

* The bases are orientated towards the centre of the helix, protecting them from reacting with other chemicals.

* Bases on one strand have **specific base pairing** with bases on the other strand.

> **Adenine always pairs with thymine.**
> **Guanine always pairs with cytosine.**

* Bases are joined by weak **hydrogen bonds**, but there are so many between the strands that together they make DNA a **stable polynucleotide**.

* The DNA helix is further coiled to produce a super helix, providing a compact store of genetic information.

> In almost any question concerning DNA or RNA, specific base pairing needs to be referred to – and understood. You may have been taught this as complementary base pairing.

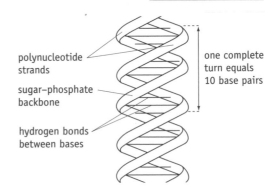

Alpha double helix

✓ *Quick check 1, 2, 3*

The role of DNA

* It carries the **genetic information** controlling the synthesis of proteins. By controlling which proteins, particularly enzymes, are produced in a cell, DNA controls the development, structure and function of a cell.

* It is capable of **self-replication**, which is essential for increases in cell number during growth, reproduction (asexual and sexual), and passing genetic information to the next generation.

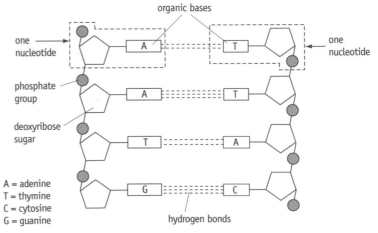

A = adenine
T = thymine
C = cytosine
G = guanine

Structure of DNA

- Although DNA is a relatively stable molecule, alterations in the genetic information (**mutations**) can occur, providing **variation** – the basis for evolution via natural selection.

You will need to know similarities and differences in structure between DNA and RNA.

B

HB

RNA

Structure of RNA

The structure of RNA differs from DNA in that:

- the pentose is **ribose** not deoxyribose;
- the bases found in RNA are adenine, guanine, cytosine and **uracil**; uracil replaces thymine and base-pairs with adenine.
- it is **single-stranded**, not double-stranded (although the single strand can fold back on itself);
- it is **shorter** than DNA, with a lower molecular mass.

DNA is much too large to be soluble, or to pass through the pores in the nuclear envelope – it stays in the nucleus. mRNA and tRNA are small enough to be soluble and move around the cytoplasm of the cell.

Types of RNA

There are three types of RNA. Ribosomal RNA is a structural component of ribosomes. Messenger RNA (mRNA) and transfer RNA (tRNA) have important roles in protein synthesis.

Messenger RNA (mRNA)

- This is a single polynucleotide strand formed in the nucleus during **transcription**, using a specific section of a DNA molecule (a **gene**) as a blueprint or template.
- mRNA carries a 'copy' of the genetic information of the gene to ribosomes in the cytoplasm.
- The mRNA is used in **translation** to determine the sequence of amino acids in a protein (its primary structure).

Transfer RNA (tRNA)

- A tRNA molecule is a single strand, folded into a 'clover leaf' shape.
- There are different types of tRNA molecule in the cytoplasm, each with a binding site for the attachment of a **specific amino acid**.
- During protein synthesis each tRNA molecule carries its specific amino acid to mRNA on a ribosome.
- A specific sequence of three bases on the molecule is known as the **anticodon**. This sequence is complementary to a **codon** on the mRNA.

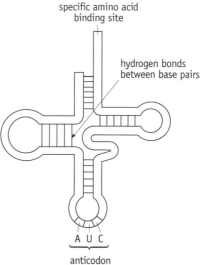

specific amino acid binding site

hydrogen bonds between base pairs

A U C

anticodon
messenger RNA binding site

Structure of tRNA

✓ *Quick check 4*

✓ *Quick check 4*

? *Quick check questions*

1 Give the components of a nucleotide.
2 Name the four organic bases present in DNA.
3 How are the bases joined together in DNA?
4 Give two ways in which the structure of: (**a**) RNA differs from DNA; (**b**) tRNA differs from mRNA.

DNA replication and the genetic code

DNA replication occurs as part of the process of cell division, and is essential for the growth and reproduction of organisms.

The semi-conservative mechanism of DNA replication

When DNA replicates:

- its double helix uncoils (unzips) into two separate strands as hydrogen bonds between the polynucleotide strands are broken;
- each strand acts as a **template** for the formation of a new complementary strand;
- **nucleotides** bind to each template strand by **specific base pairing**;
- adenine pairs with thymine and cytosine pairs with guanine;
- the nucleotides are joined together by the enzyme **DNA polymerase** to form a **polynucleotide strand**;
- the two new DNA molecules are **identical** to each other and to the original DNA;
- each newly formed DNA molecule contains one of the original polynucleotide strands, hence the term **semi-conservative replication**.

parental DNA

first generation – each DNA molecule contains one of the parental polynucleotide strands

individual DNA nucleotides join to parental DNA strands according to their specific base pairings

new polynucleotide DNA strands

new DNA strands

original parental DNA strand

parental DNA strands separate – each acts as a template for formation of identical DNA molecules

original parental DNA strand

two identical DNA molecules

Semi-conservative mechanism of DNA replication

✓ *Quick check 1, 2*

The gene

Genes are lengths of DNA which contain coded genetic information that determines the nature and development of organisms.

- A gene is the **sequence of (nucleotide) bases of DNA** coding for the production of a **specific polypeptide** by determining the sequence of amino acids (primary structure).
- Genes are located along **chromosomes** – thread-like structures consisting of DNA and protein.
- A gene can exist in different forms, called **alleles**, that code for alternative versions of the same characteristic.
- Alleles of a particular gene are located in the same relative position (**locus**) on homologous chromosomes.
- Chromosomes in a homologous pair carry the same genes, but not necessarily the same alleles of those genes.

Gene and allele do not mean the same thing – you need to be careful which you refer to.

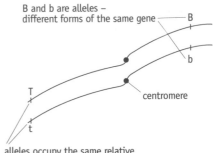

B and b are alleles – different forms of the same gene

B

b

centromere

T

t

alleles occupy the same relative position (locus) on homologous chromosomes

A pair of homologous chromosomes

✓ *Quick check 3, 4*

The genetic code

DNA carries **genetic information** that determines the sequence of amino acids in proteins. The **genetic code** is founded on sequences of (**nucleotide**) **bases**.

- A sequence of three (nucleotide) bases is called a **base triplet**.
- The presence of four different nucleotides in DNA and RNA means there are 64 (4^3) possible base triplets.
- These code for the 20 commonly occurring amino acids in living organisms.
- The base triplets of mRNA are known as **codons**.
- The genetic code is **degenerate** as some amino acids are coded for by more than one codon, e.g. six different codons code for the amino acid arginine.
- Some codons are used for 'punctuation'; for example, a 'stop' or nonsense codon codes for the end of a particular gene.

The genetic code is **non-overlapping**: each triplet is read in sequence as if it were a single word, separate from the triplet before and after it. For example:

- The sequence of bases **ATCGGCATT** is read as three non-overlapping triplets – **ATC GGC ATT**;
- If the code was overlapping, this could be read differently by starting at a different base. For example, starting at the first **T** to give – **TCG GCA TT?**

> The genetic code is universal, with all organisms using the same triplets of nucleotide bases to code for the same amino acids.

✓ *Quick check 5*

Introns and exons

There are large amounts of DNA which do not carry genetic information.

- These 'junk' lengths of DNA base sequences are called **introns**.
- DNA base sequences that do carry useful genetic information are called **exons**.
- A gene may contain many introns and these are removed during the process that uses the information on DNA to make a polypeptide.

? Quick check questions

1 Explain what is meant by semi-conservative replication.
2 What is the role of the enzyme DNA polymerase during DNA replication?
3 What is an allele?
4 Give a precise definition of a gene.
5 Explain what is meant by a degenerate genetic code.

DNA and protein synthesis

Protein synthesis is divided into two processes: **transcription** and **translation**. Transcription occurs in the **nucleus** and involves '**rewriting**' (transcribing) part of the DNA code into a strand of messenger RNA. Translation occurs in the cytoplasm and involves ribosomes synthesising proteins using the information provided by messenger RNA.

Transcription

During transcription:

- the section of the DNA molecule (a gene) uncoils and the two polynucleotide strands separate as hydrogen bonds are broken;

- one strand is the template or **sense strand**;

- RNA nucleotides line up alongside the DNA nucleotide bases on the template strand by specific (complementary) base pairing, with uracil (RNA) pairing with adenine (DNA);

- the enzyme **RNA polymerase** joins the RNA nucleotides together to form a strand of mRNA which contains introns and exons;

- introns are removed before the mRNA strand is translated.

The mRNA strand leaves the nucleus through a nuclear pore and attaches to a ribosome in the cytoplasm, where translation occurs. The strands of DNA in the nucleus will re-coil when sufficient mRNA has been produced.

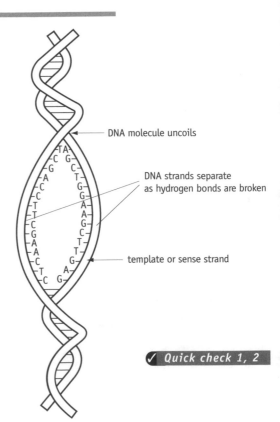

DNA molecule uncoils

DNA strands separate as hydrogen bonds are broken

template or sense strand

✓ *Quick check 1, 2*

Translation

During translation, the sequence of codons on the mRNA strand determines the sequence of amino acids in a polypeptide. Translation is carried out at ribosomes in the cytoplasm. In the cytoplasm there is a specific type of tRNA for each of the 20 amino acids. Each tRNA molecule has three exposed bases known as an **anticodon**.

- An mRNA molecule attaches to a ribosome.

- A tRNA molecule with the complementary anticodon binds to the first codon on the mRNA strand, bringing its specific amino acid.

Transcription – DNA strands separate

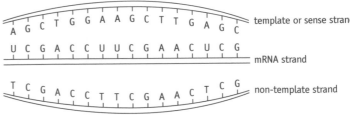

template or sense strand

mRNA strand

non-template strand

Transcription – formation of RNA

- The **anticodon** binds to the **codon** by **specific (complementary) base pairing**.
- Another tRNA then binds to the second codon on the mRNA strand.
- The amino acid on the first tRNA molecule is joined to the amino acid on the second tRNA molecule by a peptide bond.
- This requires ATP and the action of an enzyme.
- The first tRNA molecule then moves away from the ribosome leaving its amino acid behind, and the mRNA moves over the ribosome by one codon.

As the mRNA strand has been transcribed from the DNA template strand, it is the sequence of DNA (nucleotide) bases that ultimately determines which specific polypeptide is produced.

B

HB

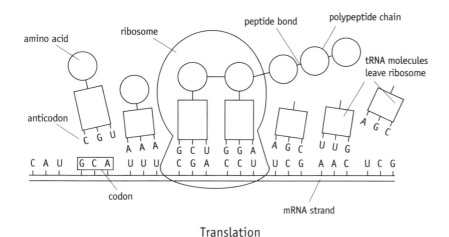

Translation

- This process continues along the mRNA strand until all the codons have been 'read' and the primary structure of the polypeptide has been produced.
- The polypeptide folds itself into its secondary and tertiary structure.
- The sequence of amino acids in this polypeptide has been determined by the sequence of codons on the mRNA strand.

✓ *Quick check 3, 4*

? Quick check questions

1 Give the RNA sequence which would be complementary to the following DNA bases: T T G C G A C G T G G C A

2 Give two ways in which DNA replication and transcription differ.

3 Name the product of translation.

4 Describe the role of tRNA molecules during translation.

B
HB

Recombinant DNA

In genetic engineering, genes may be taken from one organism and inserted into the DNA of another organism, producing **recombinant DNA**.

Genetically engineered microorganisms

Microorganisms, particularly bacteria, are widely used as recipient cells during gene transfer. The rapid reproduction of microorganisms enables a transferred gene to be copied so that a large amount of gene product can be obtained.

Obtaining genes

Genes can be removed from the DNA of one organism by using **restriction endonuclease enzymes**.

- Restriction enzymes cut the DNA to produce overlapping bases ('sticky ends').

- Each enzyme **cuts** the **DNA** at a **specific base sequence**.

Alternatively, **complementary DNA** (**cDNA**) produced from mRNA can be used as the 'foreign' gene.

- mRNA is obtained from a cell (e.g. pancreatic cell) which produces large amounts of protein.

- The enzyme **reverse transcriptase** is used to produce cDNA by the reverse process of normal transcription.

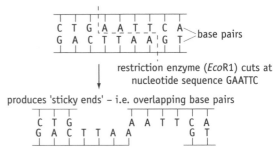

Action of a restriction endonuclease enzyme

Transfer of genes

A **vector** is usually required to transfer the 'foreign' gene into bacterial cells. **Plasmids** and **viruses** are commonly used as vectors.

- Plasmids are small circular sections of DNA found in some bacteria.

- The plasmid is cut using the **same restriction endonuclease** enzyme used to remove the 'foreign' gene.

- The plasmid DNA and the 'foreign' DNA (gene) are then joined together using a **ligase** enzyme that fits together the overlapping base pairs ('sticky ends').

- The plasmid and the foreign DNA are referred to as **recombinant plasmid**.

- These plasmid vectors are added to a culture of bacteria, some of which take up the recombinant plasmid by a process called **transformation.**

Viruses (phages) can be also used as vectors.

- Viral vectors infect the bacteria, inserting the 'foreign' gene into the cell.

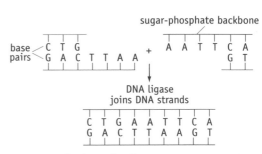

Action of a ligase enzyme

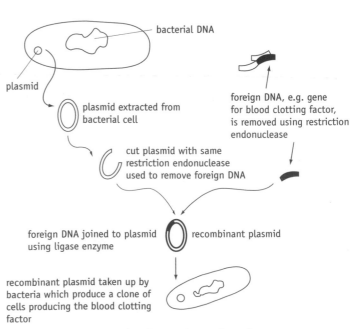

Example of genetic engineering

✓ *Quick check 1, 2*

Obtaining gene products

- Any bacteria taking up the transferred gene will replicate it during cell division, producing a clone of genetically engineered bacteria.

- The bacterial cells are cultured and produce the gene product in large amounts (e.g. insulin, blood clotting factors).

Genetic markers in plasmids

These are genes already present in the plasmid and which therefore enable genetically engineered bacteria to be detected for subsequent culturing. Genetic markers include genes in the plasmid which confer **antibiotic resistance**, e.g. ampicillin resistance.

- Initially the recombinant plasmids are added to bacteria on solid nutrient agar medium (**master plate**).

- After colonies have grown they are transferred to **replica plates** containing the antibiotic, ampicillin.

- Only bacteria containing the recombinant plasmid will grow as they have the ampicillin resistance gene (genetic marker) and are not destroyed by the antibiotic.

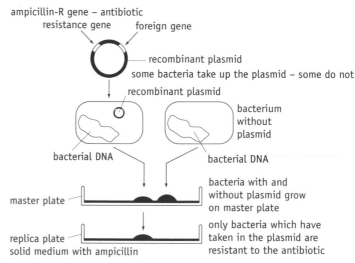

Introduction and detection of recombinant plasmids

✓ *Quick check 3*

Evaluation of genetic engineering

Considerable debate is taking place on the risks of using genetic engineering, particularly in medicine and in food production. Some of the points raised include:

- possible transfer of foreign genes to non-target organisms, including humans, resulting in disruption of normal functions;

- irreversible process with no certainty of economic benefits;

- ethical considerations with regard to altering genetic make-up of animals;

- ecological and evolutionary consequences are unknown;

- development of new resistant species, e.g. antibiotic-resistant bacteria;

- effect of consumption of genetically engineered food containing foreign proteins;

- accidental transfer of unwanted genes by the vector, e.g. virus.

> Newspapers and science periodicals have up-to-date information on the debate concerning genetic engineering.

✓ *Quick check 4*

? Quick check questions

1 Explain the function of the following enzymes in genetic engineering:
   (a) restriction endonucleases; (b) ligase.

2 What are plasmids?

3 Explain how genetic markers enable detection of genetically engineered bacteria.

4 Describe two possible risks associated with the use of genetic engineering.

Isolation of enzymes

Enzymes are **proteins** which **control the biochemical reactions** (metabolism) of cells. Enzymes:

- are **catalysts** – they **lower activation energy** by forming an **enzyme–substrate complex**;
- allow reactions to happen under 'normal' physiological conditions of temperature and pH;
- are **specific**, because their tertiary structure fits the shape of their substrate.

This specificity is explained by the **lock and key** and **induced fit models** of enzyme action. These properties of enzymes have led to important applications in **biotechnological processes**. (See Module 1 for more details about enzymes.)

> Make sure you revise the characteristics of enzymes from Module 1.

Intracellular and extracellular enzymes

All cells make **intracellular enzymes** which work inside the cell – inside the cell surface membrane. Many cells also secrete **extracellular enzymes** – to work in the external environment. The table below shows why extracellular enzymes are more often used for biotechnological processes.

Intracellular enzymes	Extracellular enzymes
More difficult to isolate	Easier to isolate
Cells have to be broken apart to release them	No need to break cells – secreted in large amounts into medium surrounding cells
Have to be separated from cell debris and a mixture of many enzymes and other chemicals	Often secreted on their own or with a few other enzymes
Often stable only in environment inside intact cell	More stable
Purification/downstream processing difficult/expensive	Purification/downstream processing easier/cheaper

> You may well be asked to give two or three advantages of extracellular enzymes compared with intracellular enzymes.

Microorganisms such as bacteria and fungi feed saprophytically, secreting (digestive) enzymes onto their food – making them a good source of extracellular enzymes.

> ✓ *Quick check 1*

Commercial production of an enzyme

(The example given on the next page is the production of pectinase from the fungus *Aspergillus niger*. You may have studied a different example. The basic principles will be the same but there may be differences in practical detail.)

Pectin is a substance found in the cell walls of plant cells. When fruit is crushed to extract juice, pectin prevents some being released and makes the juice cloudy. **Pectinase** breaks down pectin, releasing more juice, and makes juice clear. *Aspergillus niger* is a fungus that produces pectinase.

> The name of a species should be written in italics, or underlined. The first word has a capital, the second a lower case initial.

B

- A starter culture of *Aspergillus niger* is put into a **fermenter** – where **optimum conditions** for growth are maintained.

- Under optimum conditions *Aspergillus* grows very rapidly, soon producing a large mass of fungus – which produces a lot of enzyme.

The fungus is grown in **aseptic conditions** – free from contaminating microorganisms by:

- testing the starter culture for purity – unwanted microorganisms produce unwanted substances/enzymes.

- cleaning and sterilising the fermenter – using disinfectant, detergent and hot water, and steam;

- sterilising the growth medium – using heat or UV light.

The **specific growth medium** contains:

- a nitrogen source – for protein synthesis;

- suitable vitamins and mineral ions;

- **sucrose** – as a respiratory substrate to provide energy;

- **pectin** – which causes *Aspergillus niger* to make lots of pectinase.

In most examples of the commercial production of enzymes, the substrate of the enzyme is added to stimulate the microorganism to produce more of the enzyme.

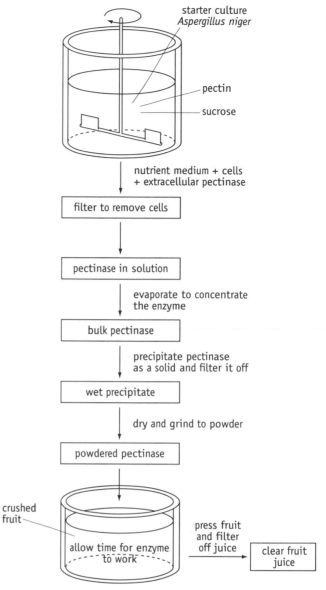

The production, downstream processing and use of pectinase from *Aspergillus niger*

✓ *Quick check 2, 3*

❓ *Quick check questions*

1 Explain why many of the enzymes used in commercial processes are extracellular enzymes from bacteria or fungi.

2 Describe the isolation and purification of a named enzyme by downstream processing.

3 A fungus was used as source of the enzyme amylase, which breaks down starch into maltose. (a) Starch was given to the fungus as its only energy source. Suggest why this was done. (b) Explain why the fungus would be grown under aseptic conditions.

B Enzymes in biotechnology

Applications of enzymes in biotechnology are linked to their characteristics and functions in living organisms (see the first section on page 48).

Analytical reagents

- Enzymes are very **specific** – they bind to their substrate only.
- This means they can be used to **identify a specific substance** in a sample.
- Enzymes have **high sensitivity** – they react with **low concentrations of substrate**, to produce measurable amounts of product(s).

Glucose oxidase is an example of an enzyme used as an analytical reagent.

test strip
– plastic

blue band
coated with
glucose oxidase
and peroxidase
– blue due to
chromagen dye

dipped in urine sample

colour change
depending on the
amount of glucose
present

compare against
colour chart to
find glucose
concentration

> Make sure you know the enzyme, its substrate and the product at each stage of the test for glucose – they have to be in the right order and place.

Glucose + water + oxygen $\xrightarrow{\text{glucose oxidase}}$ gluconic acid + hydrogen peroxide

Blue chromagen dye + hydrogen peroxide $\xrightarrow{\text{peroxidase}}$ green to brown chromagen dye + water

Testing urine for glucose

✓ *Quick check 1*

Thermostability

Most of the enzymes that you study have optimum temperatures between 20° and 40°C. **Thermophilous** microorganisms live in very hot environments, such as hot springs/volcanic vents. Their enzymes have optimum temperatures as high as 115°C; they are **thermostable**.

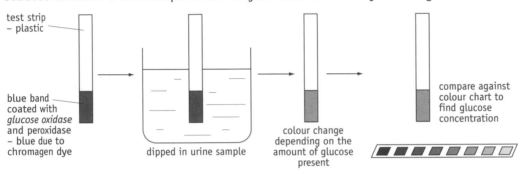

cheap starch paste from corn (maize)

amylase – works between 90° and 100°C

short chains of glucose molecules

amyloglucosidase – optimum temperature of about 55°C

glucose syrup

glucose isomerase – working at 60°C

fructose rich syrup – much sweeter than glucose syrup

> Fructose syrup is added to diet/ slimming food to make it taste sweet enough to suite customers' taste. Because it is so sweet, less has to be added, hence less carbohydrate and fewer calories/joules.

Thermostable enzymes from microorganisms used in fructose production for diet foods

Thermostable enzymes are useful in industrial processes because:

- reactions occur at a **faster rate at higher temperatures** – producing more product, more quickly;
- the temperature of the fermenter/reactor does not have to be monitored and controlled as closely as with more 'delicate' enzymes – which saves money.

✓ *Quick check 2*

Immobilised enzymes

Obtaining pure enzyme from a microorganism in large amounts is expensive. To make their use economically viable on an industrial scale, enzymes have to be re-used many times. As catalysts, enzymes can be used very many times – if they are not lost with the products of the reaction. Enzymes can be **immobilised**, allowing them to be separated easily from products (and re-used). For example:

- **cross-linkage** – enzyme held together by a cross-linking agent;
- **entrapment** – enzyme held in a mesh or capsule of an inert material;
- **adsorption** – enzyme attached to the outside of an inert material.

Immobilisation **reduces the activity of enzymes** but often makes them **more heat- and pH-stable**.

Immobilised enzymes can be used in **continuous fermenters** (reactors). The advantages involve the **control** of the enzyme reaction in terms of:

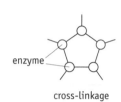
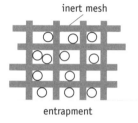

Immobilisation of enzymes

- maintenance of optimum conditions for the enzyme, such as temperature and pH – giving a high rate of reaction;
- substrate added continuously at the optimum rate and concentration – giving a high rate of reaction;
- product(s) can be removed continuously – giving high productivity;
- these are pure and free of enzyme – reducing costs of purification.

The fermenter can also run for long periods before it has to be stopped, cleaned and re-started – saving money on labour and cleaning.

Immobilised protease can be used this way to make amino acids from protein.

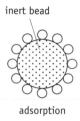 Immobilisation changes the environment of an enzyme. That changes its tertiary structure, which is certain to change its activity.

The AQA specification refers to separation of products from reactants – so reference to encapsulated enzymes in washing powders is not likely to be relevant.

✓ *Quick check 3, 4*

Quick check questions

1 Describe one example of the use of an isolated enzyme as an analytical reagent.

2 Use one example to explain the importance of thermostable enzymes in an industrial process.

3 Explain the advantages of using immobilised enzymes in a biotechnological process.

4 Suggest the factors that would be considered when trying to decide whether or not to use an enzyme in a biotechnological process.

B Immunology and blood groups

Forensic examinations of blood samples identify people's blood groups, in connection with evidence from scenes of crimes or to establish paternity.

Immunology

The body defends itself against invading foreign organisms, cells and large organic molecules. This **immune response** involves **lymphocytes**.

Antigen and antibody

Antigens are:

- **proteins**, **polysaccharides** or **glycoproteins** causing an **immune response**;
- usually on cell walls or plasma membranes of cells.

An **antibody** is:

- a **protein** produced by a **B-lymphocyte**, that **binds to a specific antigen;**
- specific because of its unique **tertiary structure** – its 3-dimensional shape.

Binding of the antibody leads to destruction of the antigen (bearing cell).

Immunological response of B-lymphocytes

- B-lymphocytes are one of several types of white blood cell.
- As each B-lymphocyte differentiates/specialises, it produces its **own specific antibody,** inserted in its plasma membrane/cell surface membrane.

If a B-lymphocyte meets its specific antigen, the following happens:

- The antigen binds to a **receptor site** on the **antibody** in the membrane.
- It does this because the antigen fits the shape of the receptor.
- Binding makes the B-lymphocyte start **secreting antibody** into the blood.
- It also starts **dividing by mitosis**, producing a **clone** (clonal expansion) of genetically identical B-lymphocytes, called **plasma cells.**
- These plasma cells secrete the same antibody – destroying the antigen.

When all the antigen has been destroyed:

- most of the plasma cells die, but some plasma cells remain as **memory cells**, usually for many years;
- memory cells give a **fast response** to another invasion by the **same antigen** – fast enough to prevent any harm by the antigen.

> As with any protein, the function of an antibody is linked to its tertiary structure – its 3-D shape. You should look at questions with this in mind.

> ✓ *Quick check 1, 2*

> If a B-lymphocyte does not meet its antigen, it dies. The body does not produce a specific B-lymphocyte in response to invasion by an antigen. The body produces an almost infinite number of different B-lymphocytes, so that hopefully there will be one that binds to any antigen that invades.

> See the diagram on page 80 – antibody production by a B-lymphocyte.

> ✓ *Quick check 3, 4*

ABO blood groups

ABO blood groups are determined by the **antigens** on the **cell surface membranes of red blood cells.**

- **Type A** has red blood cells with **A antigens** (agglutinogen) on their plasma membranes, and produce **anti-B antibody** (agglutin).
- **Type B** have **B antigens** and **anti-A antibody**.

A antigen   red cell   B antigen   A and B antigens   no A or B antigens

type A   type B   type AB   type O

with anti-B antibody in blood plasma

with anti-A antibody in blood plasma

with no anti-A or anti-B antibody in blood plasma

with anti-A and anti-B antibody in blood plasma

antibody

ABO blood groups

- **Type AB** have **both antigens** but **no antibodies** against A or B.
- **Type O** have **no A or B antigens** but **both anti-A and anti-B antibodies**.

Agglutination

If type A blood is mixed with type B blood and looked at under a microscope, the red blood cells will be stuck together in clumps (**agglutinated**) because:

- anti-A antibodies stick type A cells together;
- anti-B antibodies stick type B together.

This means that blood groups need to be matched for blood transfusions.

- **Type O** blood can be given to anyone in an emergency, because it has no antigens to lead to agglutination – **universal donor.**
- Type AB can receive blood from anyone, because there is no anti-A or anti-B to cause agglutination – **universal recipient.**

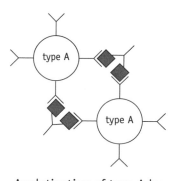

type A

type A

Agglutination of type A by anti-A antibody

✓ *Quick check 5*

? *Quick check questions*

1 Explain what is meant by: (**a**) an antigen; (**b**) an antibody.

2 Explain why an antibody binds only to its specific antigen.

3 Describe how B-lymphocytes respond to a foreign antigen.

4 A child was injected with a vaccine containing an antigen from a disease organism. Suggest how this can lead to the child being able to fight off the disease in future.

5 Explain why giving type A blood to someone with type O is dangerous.

B

Genetic fingerprints and the polymerase reaction

It is possible to identify individuals from very small samples of DNA, by looking for unique patterns in their DNA.

Genetic fingerprinting

We all have DNA making up our genes, but about 90% of DNA has no function.

- This **non-functional DNA** consists of **short sequences of bases.**
- These are often **repeated end-to-end** many times – forming a **VNTR** (variable number of tandem repeats).
- There are many different VNTRs.
- The **position** of these VNTRs in DNA is **different for each person** – unless you have an identical twin.
- A person can be identified from a sample of their DNA by the position of the VNTRs.
- DNA in white blood cells in a **blood sample** can be extracted and analysed. (Remember – red blood cells have no nuclear DNA.)
- The **polymerase chain reaction** can be used to increase the amount of DNA available for analysis (see opposite page).

> ▶ The non-functional DNA is very variable because of accumulated mutations. Mutations are normally harmful to an organism but have no effect in non-functional DNA.

✓ *Quick check 1*

Use of restriction enzymes

- DNA is cut into small fragments by **restriction (endonuclease) enzymes**.
- Each restriction enzyme is **specific** – it cuts DNA wherever a specific base sequence occurs.
- By using the same restriction enzymes, each person's DNA will be cut into a **unique set of fragments**, containing a unique pattern of VNTRs.

Separation of DNA fragments

- DNA fragments are separated by **gel electrophoresis**.
- The mixture of fragments is placed in a well at the top of the gel.
- Smaller DNA fragments move further in the gel when an electrical potential is applied (see diagram).
- There is a **unique pattern of fragments** (with VNTRs) for each person.

Radioactive DNA probes

- A **nylon membrane** placed on the gel picks up a sample of the DNA fragments – still in order.
- DNA fragments are treated to separate their double helices into **single strands**.
- Radioactive **DNA probes** are then put on the nylon.

```
C  T  G  A  A  T  T  C  A
G  A  C  T  T  A  A  G  T
```
> base pairs

restriction enzyme (*Eco*R1) cuts at nucleotide sequence GAATTC

produces 'sticky ends' – i.e. overlapping base pairs

```
C  T  G            A  A  T  T  C  A
G  A  C  T  T  A  A            G  T
```

Action of a restriction enzyme

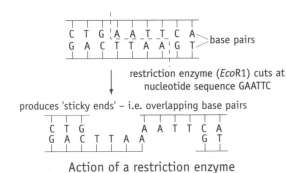

anode +ve

electric potential applied – DNA fragments (–ve charge) move towards +ve terminal

cathode –ve

gel through which DNA fragments move

smallest DNA fragments move furthest

DNA fragments placed in well

different DNA sample in each well – all cut with same endonucleases

Gel electrophoresis

✓ *Quick check 2*

B

- Each probe is a single nucleotide strand with a base sequence complementary to a particular VNTR.
- The sugar-phosphate backbone of the probe is **radioactively labelled**.
- Each DNA probe binds to its VNTR by **complementary base pairing**.
- X-ray film is placed on the nylon and 'fogs' where radioactive DNA probes are present – giving a black band.
- There is a **unique set of bands** for each person – a **genetic fingerprint**.

The diagram shows how DNA fingerprints can be used to prove relationships.

✓ *Quick check 3*

DNA extracted from blood cells (or other body tissues)

restriction enzymes added

fragments of DNA

electrophoresis through a gel

the short lengths of DNA travel further than the long lengths; the bands are not visible at this stage

heated to separate double-stranded DNA; copy transferred to nylon membrane; radioactive probe applied

X-ray film

nylon membrane

then autoradiograph made

only the DNA bands that hybridised with the radioactive probe affect the X-ray film

M C C F

both children (C) share bands with each parent (M, F), showing they are related

DNA probes and genetic fingerprints

The polymerase chain reaction (PCR)

This technique produces large quantities of identical DNA from a small sample – **DNA amplification**. Small samples of DNA are obtained at crime scenes, or for paternity cases.

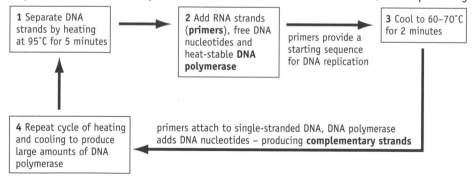

1 Separate DNA strands by heating at 95°C for 5 minutes

2 Add RNA strands (**primers**), free DNA nucleotides and heat-stable **DNA polymerase**

primers provide a starting sequence for DNA replication

3 Cool to 60–70°C for 2 minutes

4 Repeat cycle of heating and cooling to produce large amounts of DNA polymerase

primers attach to single-stranded DNA, DNA polymerase adds DNA nucleotides – producing **complementary strands**

Polymerase chain reaction

❶ DNA polymerase is obtained from bacteria living in volcanic vents and is not denatured at the high temperatures used in PCR.

✓ *Quick check 4*

❓ *Quick check questions*

1 Explain how each person's DNA is unique.

2 Describe and explain how a unique pattern of DNA fragments can be obtained from a person's DNA for genetic fingerprinting.

3 Explain how a genetic fingerprint is produced from the result of gel electrophoresis of DNA fragments.

4 **a** Explain how you would increase the amount of DNA for an analysis.

b A suspect says that the fact that a sample from a crime shows the same genetic fingerprint as himself is purely due to chance. Explain why this is most unlikely.

B Adaptations of cereals

Cereals are cultivated **grasses**, producing **grains** that form an important part of the human diet. Each species has structural and physiological adaptations to grow in a particular environment. The choice of which cereal species to grow depends on matching a cereal species' adaptations to the local environment.

> Read questions carefully – are they asking for a structural or a physiological adaptation?

Rice

Rice is a **swamp plant**, grown in flooded paddy fields. The water-logged soil/mud has no air spaces containing oxygen. Most other plants cannot survive in these conditions, because without oxygen for aerobic respiration their roots die.

Rice plants are adapted for their environment:

- **structurally** by having **aerenchyma** – large air spaces in the leaves, stem and roots, allowing oxygen to diffuse from the air to the roots;

- **physiologically** by **tolerating ethanol** from anaerobic respiration – so their roots get energy from anaerobic respiration without poisoning themselves.

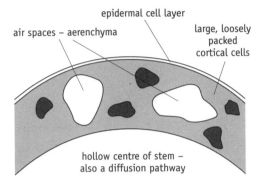

epidermal cell layer

air spaces – aerenchyma

large, loosely packed cortical cells

hollow centre of stem – also a diffusion pathway

Aerenchyma in a rice stem

> ✓ *Quick check 1*

Maize

Maize (corn) is a **tropical plant**, grown where light intensity and daytime temperature are very high. These conditions should favour photosynthesis, but:

- high temperatures increase water loss by evaporation (transpiration) – leading to the closing of stomata;

- plants compete for carbon dioxide – lowering the carbon dioxide concentration near close-packed plants;

- these two factors lead to low carbon dioxide concentrations in leaves;

- closing stomata causes a build-up of oxygen from photosynthesis in the leaves;

- oxygen inhibits the enzyme used in photosynthesis to 'fix' carbon dioxide into a 3-carbon acid (GP), later converted into a sugar – so oxygen reduces photosynthetic yield.

Maize plants are adapted for their environment **physiologically** by having:

- a different enzyme (PEP carboxylase) for 'fixing' carbon dioxide in leaf mesophyll cells –with a very high affinity for carbon dioxide, allowing efficient photosynthesis at low carbon dioxide concentrations;

- 'fixation' of carbon dioxide into a 4-carbon acid (malate) – used to 'carry' carbon dioxide to bundle sheath cells, where it is used for the normal photosynthetic pathway.

> Do not try to learn a lot of unnecessary biochemical details of photosynthesis – they are not needed and will not get credit in an exam.

Maize plants are adapted for their environment **structurally** by having:

- two types of chloroplast – those in the mesophyll cells and those in bundle sheath cells;

- bundle sheath cells arranged around the vascular bundles (veins) of leaves;

- the ability to roll their leaves to reduce the number of stomata exposed to the atmosphere.

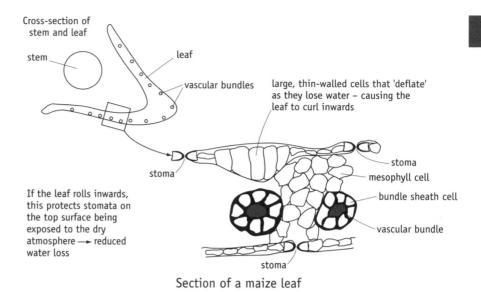

If the leaf rolls inwards, this protects stomata on the top surface being exposed to the dry atmosphere → reduced water loss

Section of a maize leaf

✓ *Quick check 2*

◐ Remember that you should use water potentials. Water vapour diffuses from a higher to a lower (more negative) water potential.

Sorghum

Sorghum is a **xerophyte**, able to grow in hot, dry conditions which kill many other species because they cannot get enough water, they lose too much water by evaporation (transpiration), or their enzymes are denatured.

Sorghum plants are adapted for their environment **structurally** by having:

- an **extensive, deep root system** – reaching the water available;
- a **thick waxy cuticle** – reducing water loss by evaporation from leaves;
- **small numbers of sunken stomata** – fewer openings out of which water vapour can diffuse (transpiration) – sunken so that water vapour builds up near the opening, reducing the water potential gradient and slowing diffusion;
- the ability to roll their leaves to reduce the number of stomata exposed to the atmosphere.

Sorghum plants are adapted **physiologically** by having:

- adult and embryo plants that can tolerate heat – their enzymes are not denatured.

◐ The cells that 'deflate' (lose turgor) to make these leaves roll are said to be 'bulliform'.

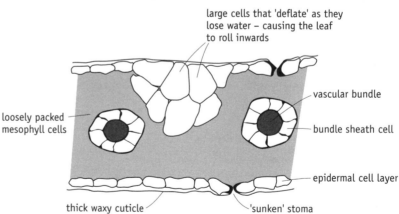

Section through a *Sorghum* leaf

? *Quick check questions*

1 Describe one structural and one physiological adaptation of rice that allows it to grow in swamp conditions.

2 Explain how the specialised method of photosynthesis in maize increases efficiency at high temperatures and low carbon dioxide concentrations.

3 Explain how having a thick waxy cuticle and fewer, sunken stomata enables *Sorghum* to grow in hot, dry conditions.

✓ *Quick check 3*

B

The abiotic environment

The **abiotic environment** is the physical/non-living part of the environment (as opposed to biotic or 'living'). Abiotic factors are not usually at the optimum for growth of crops; one or more will be a **limiting factor**. Humans manipulate abiotic factors to increase the **productivity** of plants – giving **greater yields and profits**.

Limiting factors

The limiting factors for photosynthesis are **light intensity**, **temperature** and **carbon dioxide concentration**.

From the graph you can see that the rate of photosynthesis increases:

- initially as light intensity increases – light is the limiting factor, but the curves level off as other factors become limiting;
- with a rise in temperature – when light is not limiting and carbon dioxide concentration is constant, as seen in the bottom two curves;
- with the concentration of carbon dioxide – when light and temperature are not limiting, as seen in the top two curves.

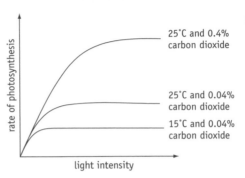

Effects of light intensity, temperature and carbon dioxide concentration on the rate of photosynthesis

▶ You might be given a graph similar to this in a question and be asked to comment on the effects of these factors. Look for the effects of each factor and then effects of combinations of changes in factors.

The highest rate of photosynthesis occurs when all factors are at their optimum. Temperatures above the optimum denature enzymes used in photosynthesis.

Commercial glasshouses

A glasshouse is a closed environment (unlike a field), where growers try to maintain an optimum abiotic environment for plants.

- Artificial lighting can increase light intensity to its optimum.
- Heaters raise temperature to its optimum.
- Heaters burning fossil fuel (such as gas) increase the carbon dioxide concentration in the glasshouse.

▶ Growers make commercial decisions. For example, it is only worth spending money on heating if there is enough light to produce extra yields that more than pay for the heating.

Fertilisers

Plants take up mineral ions from the soil. Three important ones are:

- **phosphate** ions – used to make nucleic acids (DNA and RNA) and ATP;
- **potassium** ions – actively transported into cells across the cell membrane;
- **nitrate** ions – used to make amino acids, proteins and nucleic acids.

In **natural ecosystems**, decomposition **recycles** mineral ions. **Crop plants** are removed at harvest, and so their **mineral ions are lost**.

✓ *Quick check 1, 2*

Yield would increase with increasing nitrate until some other factor becomes limiting

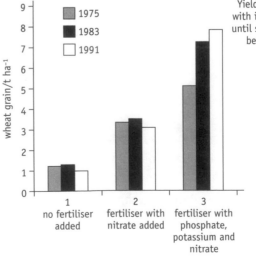

Increases in yield of wheat grain when inorganic fertiliser is added

If crops are grown year after year on the same land:

- mineral ions in the soil become depleted (very low);
- crops grow less well – leading to lower yields.

Fertilisers are used to replace lost mineral ions.

- **Organic fertilisers** (e.g. manure) – decompose on/in the soil, releasing mineral ions.

- **Inorganic fertilisers** are manufactured – consist of mineral ions and are usually sprayed onto soil in solution.

✓ *Quick check 3, 4*

	Organic fertiliser	Inorganic fertiliser
Advantages	• Slow release of mineral ions – steady supply to plants • Little leaching • Less risk of eutrophication • Add to structure of soil as humus	• Fast increase of mineral ions for periods of maximum growth • Can be added at times of maximum growth • Easy to spray on fields
Disadvantages	• Difficult to spread • Often offensive smells • Slow, gradual release of mineral ions – no response to growth periods of crop	• Repeated use damages soil structure • Sprays can be blown to other areas • Expensive • Easily leached • More risk of eutrophication

Environmental issues arise from use of fertilisers, especially inorganic ones.

- **Leaching** is the washing of mineral ions through soil – eventually getting into the water table, streams, rivers and lakes. **Nitrates** from inorganic fertilisers are very soluble and easily leached. Therefore inorganic fertiliser should not be added during/before heavy rain, or in excessive amounts.

- **Eutrophication** occurs when large amounts of nitrate and/or phosphate get into streams, rivers or lakes – by leaching or run-off of rainwater. This stimulates rapid growth of vast numbers of **microscopic algae**, which smother other water plants at the start of food chains – leading to death of many organisms. The algae die in large numbers, providing food for decomposing bacteria. Bacterial populations grow rapidly, using lots of oxygen from the water for respiration. Consequently many other organisms (including fish and invertebrates) suffocate.

Modern high-yielding crops need lots of mineral ions to produce grain, leaves, fruit, etc. This means that lots of fertiliser is needed to grow them. This is often too expensive for poor farmers in developing countries.

✓ *Quick check 5*

Remember – it is **not** the algae that use up oxygen in eutrophication.

? *Quick check questions*

1 Explain conditions in which temperature is a limiting factor in photosynthesis.

2 Explain why farmers turn up the heating in a greenhouse only on a sunny day.

3 Explain why fertilisers are added to fields.

4 Use the data in the graph opposite to answer the following questions.

   a Describe the effect of adding nitrate on crop yield.

   b Describe the effect of adding phosphate and potassium ions as well.

5 A bag of inorganic fertiliser fell into a pond. Explain the likely effects on organisms in the pond.

Pest control

The biotic (living) environment of crop plants affects productivity, yield and profits of farmers. Plants of the **same species** in a field compete for light, carbon dioxide, water, mineral ions and space; this is **intraspecific competition.** Farmers plant seeds/plants at suitable intervals in fields, irrigate (water) and add fertiliser to minimise this competition. Other species also compete with crop plants and this competition has to be minimised for good yields and profits.

> All organisms compete for the means of survival – humans are no exception! For example, humans compete with other organisms for crop plants.

Interspecific competition between weeds and crop plants

Interspecific competition occurs between **different species.** Weeds compete with crop plants for light, water, mineral ions, space and carbon dioxide.

- This is particularly important when the crop plant's seeds have just germinated and seedlings are developing root and shoot systems.

- Many weed seeds germinate earlier than crop seeds, or their seedlings grow faster.

- In both cases, the weed seedling gets its root and shoot system established first and takes water and mineral ions at the expense of the crop seedling.

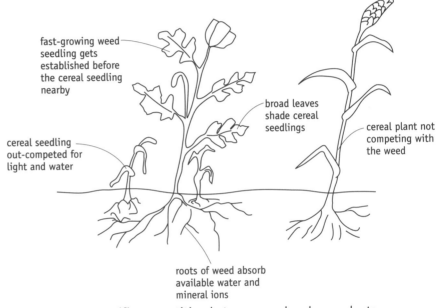

fast-growing weed seedling gets established before the cereal seedling nearby

broad leaves shade cereal seedlings

cereal plant not competing with the weed

cereal seedling out-competed for light and water

roots of weed absorb available water and mineral ions

Interspecific competition between weed and crop plants

✓ *Quick check 1*

Reduction of crop yield by insects

Insects reduce the amount of crop harvested, or damage the crop – making it unfit for use/consumption. **Insects reduce crop yield:**

- **directly** – by eating/damaging parts of the plant used by humans;

- **indirectly** – by reducing photosynthetic tissues of plants, particularly the leaves, so the plant has less energy to put into, for example, grain or fruit.

✓ *Quick check 2*

The table below gives some examples of how insects affect crop yield.

Insect pest	How yield is reduced	
	Directly	**Indirectly**
European corn borer		Larvae feed on leaves and stalks of maize
Corn earworm	Larvae eat grain of *Sorghum*	
European red mite		Feeds on leaves of apple trees
Codling moth	Larvae eat apples	
Rice water weevil		Adult beetle eats leaves of rice
Stink bug	Sucks sap from developing rice grain	

Pest control

Pesticides

Pesticide is a general term for chemicals used to kill organisms that harm crops. Pesticides are used to kill a local population of a pest organism at a particular time.

Principles of using pesticides include:

- using **specific** chemicals that kill only the pest;
- using **biodegradable** chemicals that do not persist in the environment;
- using at **appropriate times** – e.g. before critical growth period(s) of the crop; when the pest population is starting to grow rapidly;
- applying sprays in **correct weather conditions** – to avoid spray drift.

Type of pesticide	Organisms killed
Herbicide	Weeds
Insecticide	Insects
Fungicide	Fungi
Rodenticide	Rodents – rats and mice
Nematocide	Nematode worms
Molluscicide	Molluscs – slugs and snails

❶ Do not use the terms in this table as if they all mean the same thing!

Biological agents

Biological agents are organisms used for pest control. Principles include:

- using organisms that **specifically** attack the pest but not other organisms;
- aiming to **control/limit the size of pest populations** – but not kill them all;
- aiming to allow **no significant commercial damage** to the crop.

Type of biological agent	Example
Parasite of pest	Parasitic wasp laying eggs in aphids
Predator of pest	Ladybird eats aphids
Disease causing in pest	Bacterium that kills caterpillars of cabbage white butterfly
Weed eating organism	Weevil that eats purple loosestrife

✓ *Quick check 3*

Integrated systems

These use **several different methods** to control pests and aim to:

- keep pest numbers **below economically-damaging levels**;
- make it less likely that pests will adapt to one control method.

Integrated control method	Detail
Pest-resistant crops	Disease resistant; genetically modified to contain insecticide; tolerant of damage
Cultural control – farming techniques	Crop rotation; planting alternative hosts; trap plants; adjust time of planting to miss pest
Chemical	Chemical pesticides to control sudden increases in pests
Physical and mechanical	Weeding; deep ploughing; hoeing

❶ Integrated methods reduce the need to depend on large and frequent applications of pesticides – which have big environmental and practical implications (see next page).

✓ *Quick check 4*

❓ Quick check questions

1 Explain how weeds reduce the yield from crop plants.

2 Describe and explain how insects reduce crop yields: **(a)** directly; **(b)** indirectly.

3 Biological control of pests has had only limited success compared with the use of chemical pesticides. Suggest two features of biological control that make it less attractive to most farmers.

4 **a** Explain what is meant by integrated pest control.

b Explain the benefits of integrated pest control compared with using chemical pesticides alone.

B Environmental issues of pest control

Farming is about creating food chains with humans at the receiving end. The plants and animals we farm are actually part of complicated food webs, because other organisms compete with us to feed on them. Much of modern farming is about simplifying food webs by removing competition – pests. This is very important with a large and growing human population to feed.

There are a number of **environmental issues associated with pest control.**

- Many organisms that we call pests are the food of other organisms.
- Use of herbicides to kill weeds removes the food of many herbivores and the base of natural food chains.
- Use of non-specific insecticides kills many harmless insects that are the food of other organisms, e.g. some songbirds.
- They can also kill pollinating insects needed by many crops (but not cereals) for reproduction and production of seeds and fruit.
- Natural predators of pests can be killed/reduced in numbers, leading to increases in pest populations.
- Non-biodegradable pesticides persist in the environment and can be passed along food chains.
- Pesticides can get into streams, ponds, rivers and lakes and kill organisms.
- Frequent use of a pesticide may result in the evolution of resistant pests – requiring the use of more pesticides or development of new pesticides.

> In some countries as much as 25% of crop plants are eaten in one way or another by insects.

> Food chains start with plants – the producers. These are fed on by herbivores – primary consumers. Carnivores feed on the herbivores. Food chains are interconnected into complex food webs.

> Pesticides do *not* cause eutrophication.

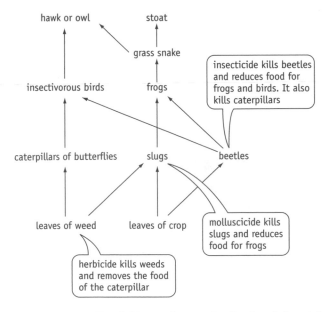

Pesticides and part of a food web involving crop and weed plants

insecticide kills beetles and reduces food for frogs and birds. It also kills caterpillars

molluscicide kills slugs and reduces food for frogs

herbicide kills weeds and removes the food of the caterpillar

✓ *Quick check 1*

Toxicity and bioaccumulation

Toxicity of a pesticide depends on the effect it has on an organism and **dosage** – the amount of pesticide in the organism.

- Small amounts may have no effect but larger doses may be harmful.

- A target organism may not die immediately – or may get a non-lethal dose.

- Pesticide will often get onto/into non-target organisms which are not killed – especially when spraying.

- Organisms containing pesticide may be eaten by organisms higher in a food chain/web.

- Pesticide may pass along food chains into top consumers, including humans. For example, insecticide sprayed on fruit can be eaten by people.

- Some pesticides are **not broken down** by organisms or excreted.

- These **accumulate in the organism** during its lifetime.

- Anything eating this organism **eats the accumulated pesticide**, which accumulates in their body.

- The result is **bioaccumulation** – accumulation of higher and higher concentrations of pesticide in organisms at each stage of a food chain.

- An example is **DDT**, an insecticide that persists in the environment (2–5 years) and is not excreted; it accumulates in fatty tissue.

- DDT is banned in this country, because it was found to be accumulating in people and had been shown to harm birds by preventing them from breeding.

> ◖ It's not always obvious how pesticides get into human food, because of complex relationships in the food webs we are part of. It isn't just what we eat but also what happened lower down the food chain.

✓ *Quick check 2*

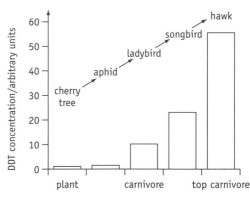

Bioaccumulation of DDT along a food chain

✓ *Quick check 3*

❓ Quick check questions

1 a Explain why farmers use pesticides.

   b Explain two ways in which the use of pesticide can damage the environment.

   c Use the information in the food web opposite to suggest how use of molluscicide to kill slugs might lead to greater damage to the crop by slugs in the future.

2 Explain what is meant by the bioaccumulation of pesticide and why it occurs.

3 Use the information in the bar chart above to calculate the percentage difference in DDT in ladybirds and hawks.

B

Reproduction and its hormonal control

Female mammals have an **oestrus cycle**, involving the production of one or more eggs. In many mammals production of eggs and mating are timed to lead to birth at the best time of year for survival of the young. The timing is set by **external environmental stimuli**. The oestrus cycle is **controlled by hormones**.

Ovarian follicle, corpus luteum and uterine endometrium

The human female is used here as an example.

- Approximately every 28 days an **ovarian follicle** starts to ripen.

- Inside the follicle, a **primary oocyte** completes the first meiotic division – forming a **secondary oocyte** (often called the egg).

- At **ovulation**, the secondary oocyte is released from the surface of the ovary and enters the **fallopian tube**, where **fertilisation** occurs.

- During fertilisation the oocyte completes the second meiotic division, forming the **egg**, which immediately **fuses with a sperm**.

- The fertilised egg – the **zygote** – divides by mitosis to form an **embryo**.

- This **implants** in the **endometrium** – the lining of the **uterus**.

- The endometrium develops during the cycle, to be ready for implantation.

- If fertilisation does not occur, the lining of the uterus is shed and re-grows.

> Not all mammals have the same oestrus cycle as humans. Many only have one cycle in a year. Some only undergo oestrus if certain environmental stimuli are present – which often affects zoo breeding programmes.

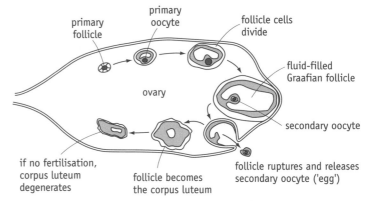

Development of an ovarian follicle and corpus luteum

> Look at the time-scale you are given on a graph/chart in a question – don't just answer from memory. The example you are given might not be human.

✓ *Quick check 1, 2*

Changes in the uterine endometrium during the oestrus cycle

Hormonal control of the female sexual cycle

Hormones are chemical messengers produced by endocrine glands, released into the blood and causing a response in their target cells.

- Hormones controlling the oestrus cycle are produced by the **pituitary gland** in the brain and by the **ovary**.

- The **hypothalamus** in the brain has links to the pituitary gland and controls the release of the hormones that control the oestrus cycle.

- This gives a link between **external environmental stimuli** (processed through the nervous system), the oestrus cycle and reproduction.

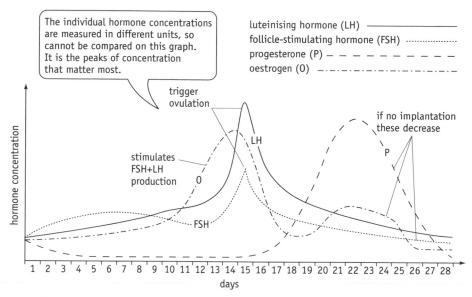

The individual hormone concentrations are measured in different units, so cannot be compared on this graph. It is the peaks of concentration that matter most.

luteinising hormone (LH)
follicle-stimulating hormone (FSH)
progesterone (P)
oestrogen (O)

Changes in concentrations of hormones during the human oestrus cycle

The main hormones controlling the oestrus cycle are **FSH** (follicle stimulating hormone), **LH** (luteinising hormone), **oestrogen and progesterone**.

- **FSH** stimulates the **development of a primary follicle** and **secretion of oestrogen** by the ovary.

- **Oestrogen** causes the development of the lining of the uterus, the **endometrium**, ready for implantation.

- It also **inhibits** the release of FSH from the pituitary gland by **negative feedback** – the release of FSH leads eventually to the inhibition of its own release.

- This prevents the development of any more follicles.

- When oestrogen levels peak, this causes a sudden **surge** in production of **FSH and LH** – leading to **ovulation**.

- After ovulation the empty follicle develops into the **corpus luteum**, which produces the hormone **progesterone**.

- **Progesterone and oestrogen inhibit FSH** production.

- If no embryo implants, the corpus luteum degenerates, progesterone production drops, inhibition of FSH is lifted and the cycle starts again.

- If the woman becomes pregnant, the corpus luteum continues to function for about 60 days and then the placenta produces progesterone – keeping up the inhibition of FSH.

> ❶ Make sure you know the names of the four hormones above, the sequence of their peaks and what each one does. Do not try to learn a lot of other detail.

> ❶ If oestrogen did not inhibit FSH, then follicles would continue to ripen and eggs be released even if the woman was pregnant. This would lead to many embryos at different stages of development trying to grow in the uterus.

✓ *Quick check 3*

❓ ## Quick check questions

1 Describe events in the ovary leading to an egg entering the fallopian tube.

2 Describe changes in the endometrium during the oestrus cycle.

3 **a** Explain how ovulation is controlled by hormones in the human female.

 b Explain the control of FSH by negative feedback.

B

Manipulation and control of reproduction

Humans control the reproduction of farm animals and their own reproduction.

Manipulation and control of animal breeding

It is useful for a farmer to know when a female animal is **in oestrus** – when she is fertile and **likely to become pregnant** if mated.

- In **cattle**, a cow in oestrus is likely to become pregnant when **artificially inseminated** – a process which is relatively expensive.

- A cow is put with a bull only when she is in oestrus – minimising risk of injury.

A cow in oestrus displays characteristic behaviour through:

- **hyper-reactivity** – restlessness;

- attempting to **mount** non-oestrus cows – who refuse;

- accepting attempts to mount by other cows (or a bull).

In **selective breeding** of cattle, farmers want to get **many offspring** from a cow with **valuable characteristics**.

- The cow is brought into oestrus by injecting a hormone, **prostaglandin**.

- Injection of **FSH** causes many follicles and eggs to develop.

- The cow is **artificially inseminated** with sperm from a superior bull.

- Many eggs are fertilised and **many embryos** implant in the uterus.

- The embryos are washed out of the uterus and **transplanted** into healthy cows with inferior characteristics.

- This protects the valuable cow from risks of pregnancy and increases the number of offspring produced.

Milk production in cows can be increased by injecting **bovine somatotrophin**, which increases the blood supply to the udder and causes an increase in the size of the udder.

Breeding behaviour in sheep can be **synchronised** so that they:

- all come into oestrus and are mated at the same time;

- produce their lambs at the same time – making it easier to plan and make arrangements for lambing, and to suit market demands.

A cow in oestrus allows mounting by other cows...

...and tries to mount other cows

Oestrus behaviour in cows

It is possible to take blood samples from animals and test for levels of hormones that indicate when oestrus is happening. This would be too expensive for any normal farmer. Farmers tend to know their animals and easily spot changes in behaviour.

✓ Quick check 1

Remember that farmers have to consider the financial costs of animal breeding. Selective breeding can increase the yield they get and the value of their herds.

Ewes in a flock can be brought into oestrus together by:

- treatment with **synthetic progesterone**;
- or introducing a castrated ram.

The central nervous system interprets information about the visual stimulus of the ram and the hypothalamus releases hormones causing the pituitary to release FSH and LH. This brings the ewes into oestrus to mate with a superior ram.

✓ *Quick check 2*

Contraceptives and control of human infertility

Hormones are used to prevent conception and pregnancy, and alternatively to treat couples who have difficulty conceiving.

The combined **contraceptive pill**:

- contains synthetic progesterone and oestrogen;
- causes inhibition of FSH release by negative feedback, which prevents follicle ripening;
- prevents ovulation – no egg, no pregnancy.

Treatment of human **infertility** depends on the cause of the problem, as shown in the table below.

Reason for infertility	What causes the infertility	Treatment
No ovulation	• FSH release inhibited by too much oestrogen • Not enough FSH and/or LH released	• Treat with a drug that inhibits the action of oestrogen, e.g. clomiphene • Injections of FSH and/or LH
Blocked or damaged fallopian tubes	• Egg released at ovulation but no fertilisation – or embryo cannot get to uterus	• IVF treatment: inject with FSH – causing many follicles to ripen; remove several eggs surgically from ripe follicles; fertilise with sperm in vitro and allow fertilised eggs to develop into early embryos; transplant 2/3 embryos into the mother's uterus
Low sperm count	• Not enough testosterone production	• Treatment with natural or synthetic androgens (male hormones), e.g. testosterone

✓ *Quick check 3*

? *Quick check questions*

1 a Explain how a farmer would know when a cow was ready for artificial insemination.

   b A farmer has a cow which produces a lot of milk. Explain how he could rapidly increase the number of high milk yielding cows in his herd.

2 Lambing time is usually restricted to two or three weeks on a farm. Explain how the farmer gets all the ewes to lamb at the same time.

3 A couple are unable to have a baby. The man has a normal sperm count and the woman does ovulate. Suggest one possible reason for their infertility and one possible course of treatment.

Module 2: end-of-module questions

1 a Explain two reasons why extracellular enzymes are more often used for biotechnological processes. [4]

b Describe how a named enzyme produced by a microorganism is isolated and purified by downstream processing. [3]

2 a Explain why enzymes can be used as analytical reagents. [2]

b The enzyme glucose oxidase can be used in a device called a biosensor, which can be used repeatedly to measure the concentration of glucose in a blood sample. It catalyses the reaction:

Glucose + oxygen + water → gluconic acid + hydrogen peroxide

   i Suggest one advantage of attaching the enzyme to the membrane. [2]

   ii Suggest how the biosensor is used to measure glucose concentrations. [3]

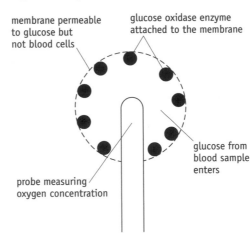

Structure of the biosensor

3 a For each of the following, describe and explain one structural adaptation it has that enables it to grow in its environment:

   i rice;

   ii sorghum. [4]

b Explain how the specialised method of photosynthesis in maize enables it to grow in a tropical climate. [4]

4 a Explain what is meant by the bioaccumulation of a pesticide. [3]

b Measurements were made of the insecticide in the bodies of animals in two food chains. The bar chart shows the insecticide in two plant eaters (the pigeon and moorhen) and two top carnivores (the sparrowhawk and the heron). The sparrowhawk eats small birds and mammals and the heron eats fish.

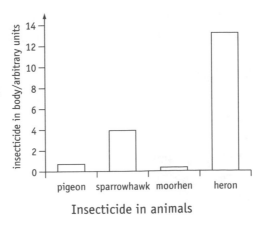

Insecticide in animals

    i Describe the differences in pesticide contents of the animals. [3]

    ii Explain the differences between the pesticide content of the
       plant eaters and the carnivores. [2]

    iii Suggest a reason for the higher pesticide content of herons. [2]

5 a A farmer was growing tomatoes in a glasshouse. He used gas-fired
     heaters to warm the glasshouse. Explain why it would be commercially
     inadvisable to use the heaters on a dull, cloudy day. [3]

  b Explain why fertilisers have to be used on fields to maintain large crop
    yields. [3]

6 a Explain the environmental impact of eutrophication. [4]

  b In 1981 a large lake was found to be suffering eutrophication.
    Efforts were made to reduce the input of nitrates into the water
    from neighbouring farmland.

key
■ algae
▨ large water plants
□ fish

Total nitrogen in organisms in each cubic metre of lake water (arbitrary units)

    i Describe the changes in nitrogen in each type of organism between
      1981 and 1991. [3]

    ii Suggest why the population of fish changed over this time. [4]

7 The diagram shows parts of the genetic fingerprints of two people who were
  suspects in a crime and a spot of blood found at the crime scene.

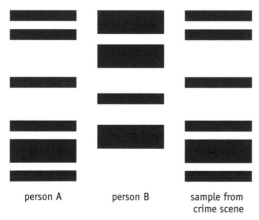

person A   person B   sample from
                                crime scene

    Genetic fingerprints

  a The spot of blood contained very little DNA. Explain how enough
    DNA was made for this analysis. [4]

  b Each dark band represents a fragment of DNA. Explain how these
    fragments were:

    i made from DNA samples; [2]

    ii made visible as dark bands. [3]

  c Suggest which suspect was more likely to be guilty. [1]

B **8** **a** The diagram shows the sequence of bases in part of an mRNA molecule.

C G U A C C U G U A A U G A U C G U C C U

  i Name the process by which mRNA is formed in the nucleus. [1]

  ii How many different amino acids could this piece of mRNA code for? [1]

  iii Give the DNA sequence which would be complementary to the first four bases in this piece of mRNA. [1]

  b The diagram shows the structure of tRNA.

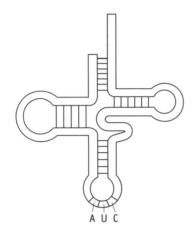

A U C

  i What term is used to describe the base triplet shown on the diagram? [1]

  ii Give two ways in which the structure of tRNA differs from mRNA. [2]

  iii Describe the role of tRNA in protein synthesis. [3]

9 The number of chromosomes in a skin cell of a mammal was found to be 44, consisting of 22 homologous pairs.

  a What is the: (i) haploid number; (ii) diploid number in this mammal? [2]

  b Explain what is meant by a homologous pair of chromosomes. [2]

  c Explain how sexual reproduction in a mammal enables a constant chromosome number to be maintained from generation to generation. [3]

  d Briefly describe the cloning of animals by splitting apart the cells of developing embryos. [3]

10 Different strains of the bacterium *Pseudomonas* are capable of breaking down various chemicals present in oil. Genetic engineering has enabled a '*super Pseudomonas*' to be produced by transferring the relevant genes into one particular strain of the bacterium.

  a Describe how genetic engineering techniques could have been used to (i) remove the relevant genes from the DNA of the *Pseudomonas* strains; (ii) insert these genes into the DNA of the '*super Pseudomonas*'. [4]

  b Suggest why the use of '*super Pseudomonas*' to remove oil slicks in the cold North Sea has been of limited success. [2]

  c Describe how genetic markers would enable '*super Pseudomonas*' to be detected amongst the other strains of the bacterium. [4]

Module 3: Pathogens and Disease

This module is broken down into seven topics: Microorganisms and disease; Blood and defence; Cell division*; The genetic code*; Gene technology*; Non-communicable diseases; Disease – diagnosis and drugs. **The topics marked * are also required by pupils taking Module 2 (Biology) and can be found on pages 36–47.**

Microorganisms and disease

- Many human diseases are caused by pathogenic organisms, including viruses, bacteria and larger parasites.
- Pathogens can reproduce rapidly, damaging host cells and sometimes producing toxins.
- Parasites have adaptations to allow them to enter the host, survive, reproduce and find a new host.

Blood and defence

- The blood defends the body against pathogens by clotting. Some lymphocytes ingest pathogens, others are involved in humoral and cellular immune responses.
- Memory cells give active immunity against secondary invasion by a pathogen.
- Active immunity follows exposure to a pathogen or vaccine. Injected antibody gives passive immunity.

Cell division B HB

- Growth and asexual reproduction depend on mitosis, producing genetically identical cells.
- In sexual reproduction gametes fuse at fertilisation. Meiosis halves the number of chromosomes in cells. Fertilisation reverses this and restores the usual chromosome number.

The genetic code B HB

- DNA carries genetic information from one generation to the next.
- The structure of DNA allows it to carry coded genetic information and replicate itself.
- A gene is a length of DNA carrying information for making a specific protein (often an enzyme).

Gene technology B HB

- Genetic engineering allows us to remove genes from one organism and insert them into another (even if it is a different species) – making it genetically engineered, or modified.
- This recombinant DNA technology raises ethical problems.

Non-communicable diseases

- Heart disease often involves obstruction of blood flow to heart muscle, due to atheroma in coronary arteries.
- Cancer is the growth of benign or malignant tumours. It is linked to mutations of genes and failure of the mechanisms that control cell division.
- The chances of developing heart disease or cancer are affected by environmental factors and lifestyle.

Disease – diagnosis and drugs

- Drugs can be used to combat disease. Beta blockers reduce high blood pressure, antibiotics kill bacteria and monoclonal antibodies target specific cells – including cancerous ones.
- Changes in the distribution and concentration of enzymes can be used to diagnose disease.
- The specificity of enzymes means they can be used in diagnostic tests.

HB Microorganisms and disease

Pathogens are one cause of disease in humans. They include viruses, bacteria and larger organisms such as the malarial parasite *Plasmodium*.

Association of microorganisms with disease

To prove that a microorganism causes a disease, Koch's postulates must be applied.

Koch's postulates

- The microorganism is present in all cases of the disease.
- It can be isolated from a diseased host and grown as a pure culture.
- The disease must be produced when a host is injected with the pure culture.
- The microorganism must be recovered from the injected host as a pure culture.

> The skin is an efficient barrier – it is waterproof, gas-proof and prevents the entry of most pathogens.

Pathogenic microorganisms **can cause** disease in their host by **damaging/killing cells** and producing **toxins** (poisons).

Pathogens have to **penetrate** one of the **body's interfaces** with the environment.

Microorganism and disease caused	Type	Interface with environment penetrated, and how	How disease is caused
Salmonella species – food poisoning	Bacterium	The gut by contaminated food and water	Toxins cause vomiting, diarrhoea and stomach cramps
Mycobacterium tuberculosis – tuberculosis	Bacterium	The lungs by droplet infection from coughs and sneezes of infected people	Damage to cells of respiratory system – causing fever and persistent cough with bloody sputum
HIV (human immunodeficiency virus) – AIDS (acquired immunodeficiency syndrome)	Virus	The skin through cuts or grazes, leading to sexual transmission – also drug-users sharing needles, and mother to baby at birth or via breast milk	Virus enters and kills white blood cells (T-lymphocytes), reducing the immune response and leaving the body open to infection by opportunistic pathogens, e.g. pneumonia bacteria

Growth of bacterial populations

✓ *Quick check 1, 2*

A host may be invaded by a few pathogenic microorganisms; for example, a few *Mycobacterium tuberculosis*. Bacterial populations can grow at an enormous rate by asexual reproduction (binary fission) and spread rapidly in the host.

Sigmoid growth curve

In culture, a bacterial population shows four growth stages/phases.

- **Lag phase** – slow growth, appropriate enzymes being produced for a new environment.
- **Log phase** – exponential growth with no limiting factors.
- **Stationary phase** – population constant due to limiting factor(s), e.g. nutrients running out, build-up of toxic wastes.
- **Decline phase** – more microorganisms are dying than being produced, some environmental factors now hostile, often food supply used up.

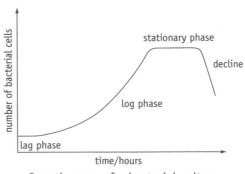

Growth curve of a bacterial culture

✓ *Quick check 3*

Sterile growth technique

Bacteria can be grown in culture **in a laboratory**. This involves **sterile technique** to **avoid contamination** of laboratory workers and contamination of cultures by unwanted microorganisms.

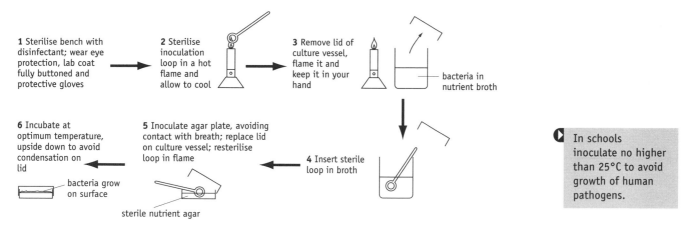

1 Sterilise bench with disinfectant; wear eye protection, lab coat fully buttoned and protective gloves

2 Sterilise inoculation loop in a hot flame and allow to cool

3 Remove lid of culture vessel, flame it and keep it in your hand

bacteria in nutrient broth

6 Incubate at optimum temperature, upside down to avoid condensation on lid

bacteria grow on surface

5 Inoculate agar plate, avoiding contact with breath; replace lid on culture vessel; resterilise loop in flame

4 Insert sterile loop in broth

sterile nutrient agar

▶ In schools inoculate no higher than 25°C to avoid growth of human pathogens.

Sterile technique in setting up a bacterial culture

Using a haemocytometer

To plot a growth curve, count the microorganisms in a liquid growth medium.

- Shake the culture at regular intervals to dipserse cells evenly.

- Take a sample of the culture liquid with a pipette, transfer it to a haemocytometer slide and put the cover slip on.

- Under a light microscope 25 large squares can be seen, each 0.2 mm down, 0.2 mm across and 0.1 mm deep – so each contains 0.004 mm^3 of culture medium and cells.

- For each time interval, count cells in 5 large squares chosen at random then divide by 5 – to give the average number of cells in 0.004 mm^3 of medium.

- To count the cells, use smaller squares inside the larger ones (see diagram).

- Multiply the result by 250 (= number of cells per mm^3), then multiply by 1000 (= cells per cm^3 of culture medium).

- Plot a graph of cells per cm^3 against time – this is a growth curve.

You can count all the cells in 5 large squares – or 80 small squares chosen at random

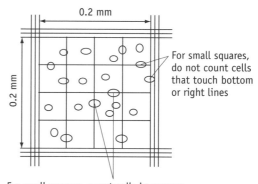

0.2 mm

0.2 mm

For small squares, do not count cells that touch bottom or right lines

For small squares, count cells in squares and those touching top and left lines

Using a haemocytometer

▶ If the bacteria were growing in a host, their numbers would fall when the host started to die, or when the host's immune system started to kill them.

? Quick check questions

1 For two named pathogens, explain how they cause disease.

2 Explain why Koch's postulates are important.

3 Describe the four main phases in the growth of a bacterial population.

HB

Parasites and parasitism (1)

Parasites live in or on their host. **Parasitism** is a one-sided relationship; the parasite gains nutrients at the host's expense. The host is damaged but usually lives for some time with the parasite.

Parasites are very specialised organisms with many adaptations for their way of life. There are advantages and disadvantages of parasitism, as shown in the table.

	Examples	**Adaptations**
Advantages – reducing energy needs	Surrounded by already-digested food	● No mouth or gut
	No need to find food	● Reduced locomotory ability ● Fewer sense organs
	Surrounded by host's body fluids	● No need to control water uptake or loss
Disadvantages – increasing energy needs	Need to find and infect new hosts	● Mobile larval stages ● Use of an intermediate host ● Produce vast numbers of offspring ● Special spines, teeth or enzymes to penetrate host's skin, gut, lungs
	Need to keep in one place in the host	● Suckers of blood flukes to attach to vessel lining ● Entry into host cells
	Need to survive hostile features of host environment	● Resistant cuticle and/or mucus to resist enzymes ● Defeating the immune system by: 'hiding' inside host cells; switching it off with secreted chemicals; having surface proteins that mimic the host's

✓ *Quick check 1*

Human immunodeficiency virus

The human immunodeficiency virus (HIV) causes AIDS – acquired immunodeficiency syndrome – which leads to death by infections/cancers that the body's immune system can no longer fight.

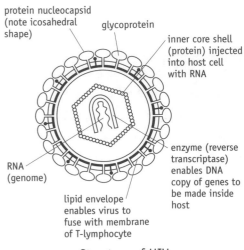

protein nucleocapsid (note icosahedral shape)

glycoprotein

inner core shell (protein) injected into host cell with RNA

RNA (genome)

enzyme (reverse transcriptase) enables DNA copy of genes to be made inside host

lipid envelope enables virus to fuse with membrane of T-lymphocyte

Structure of HIV

Life cycle of HIV

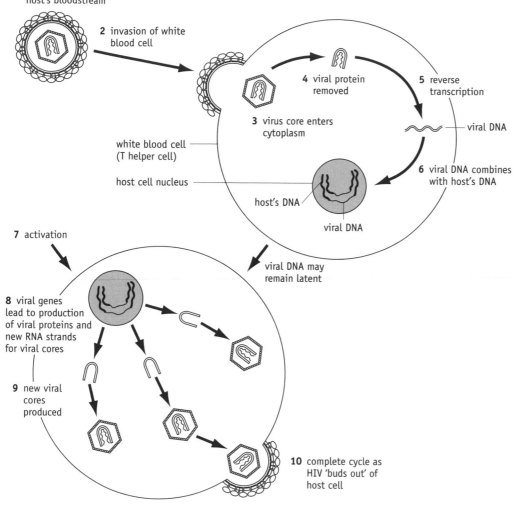

1 infection – HIV enters host's bloodstream

2 invasion of white blood cell

3 virus core enters cytoplasm

4 viral protein removed

5 reverse transcription

viral DNA

6 viral DNA combines with host's DNA

white blood cell (T helper cell)

host cell nucleus

host's DNA

viral DNA

7 activation

viral DNA may remain latent

8 viral genes lead to production of viral proteins and new RNA strands for viral cores

9 new viral cores produced

10 complete cycle as HIV 'buds out' of host cell

Life cycle of HIV

✓ *Quick check 2*

? *Quick check questions*

1 Use three examples to explain why parasites have specialised structures.

2 Explain the role of reverse transcription in the life cycle of HIV.

HB Parasites and parasitism (2)

Plasmodium

Plasmodium is a protoctist that causes malaria, a disease that kills millions of people each year. It is carried by the female *Anopheles* mosquito (the **vector**) in the salivary glands.

Life cycle of *Plasmodium*

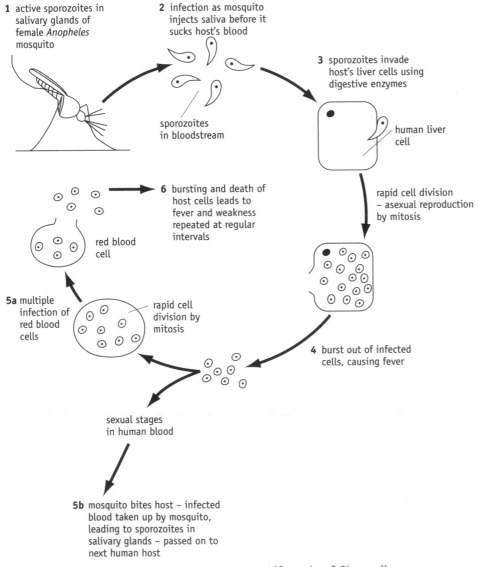

1 active sporozoites in salivary glands of female *Anopheles* mosquito

2 infection as mosquito injects saliva before it sucks host's blood

sporozoites in bloodstream

3 sporozoites invade host's liver cells using digestive enzymes

human liver cell

rapid cell division – asexual reproduction by mitosis

6 bursting and death of host cells leads to fever and weakness repeated at regular intervals

red blood cell

5a multiple infection of red blood cells

rapid cell division by mitosis

4 burst out of infected cells, causing fever

sexual stages in human blood

5b mosquito bites host – infected blood taken up by mosquito, leading to sporozoites in salivary glands – passed on to next human host

Life cycle of *Plasmodium*

- The parasite hides from the host's immune system inside host cells.
- All the nutrient molecules the parasite needs are absorbed directly from the host cell – no need to carry out digestion.
- Asexual reproduction produces vast numbers of parasites in the host.
- This makes it more likely that a mosquito vector will become infected when it feeds and carry the parasite to a new human host.

✓ *Quick check 1*

Schistosoma

This organism is a nematode worm that causes the disease bilharzia. The adult lives inside the host's blood vessels.

Life cycle of *Schistosoma*

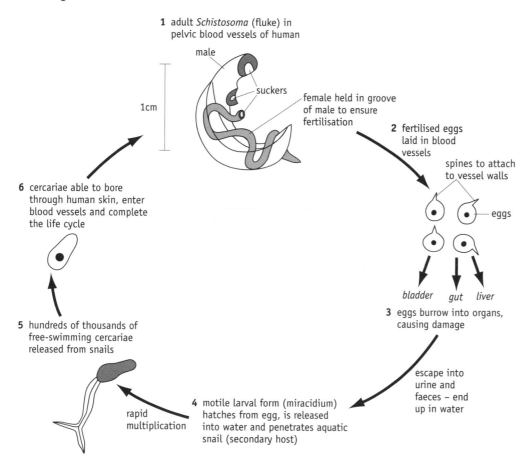

1 adult *Schistosoma* (fluke) in pelvic blood vessels of human

male

suckers

female held in groove of male to ensure fertilisation

1cm

2 fertilised eggs laid in blood vessels

spines to attach to vessel walls

eggs

6 cercariae able to bore through human skin, enter blood vessels and complete the life cycle

bladder   gut   liver

3 eggs burrow into organs, causing damage

5 hundreds of thousands of free-swimming cercariae released from snails

escape into urine and faeces – end up in water

rapid multiplication

4 motile larval form (miracidium) hatches from egg, is released into water and penetrates aquatic snail (secondary host)

Life cycle of *Schistosoma*

✓ *Quick check 2*

- Adult male has suckers to hold onto the wall of the host's blood vessels and maintain its position.

- Male holds the female to ensure fertilisation – many parasites find it difficult to find a mate and transfer gametes.

- Reproductive systems large, to produce many eggs and improve chances of infecting new host.

- Adults feed on nutrient molecules in blood plasma – their digestive system is much reduced, as are the nervous and locomotory systems.

- Adults secrete chemicals onto their surface that hide them from the host's immune system.

- Rapid asexual reproduction in snail produces many cercariae, increasing chance of infecting new human host.

? Quick check questions

1 Explain how *Plasmodium* causes disease.

2 Explain the role of the aquatic snail in the life cycle of *Schistosoma*.

HB Clotting and phagocytosis

The body has many ways to fight off invasion by pathogens. The **skin** is a **physical barrier** to infection and blood clotting seals cuts in this barrier. If pathogens enter the body, **lymphocytes** (white blood cells) of the **immune system** attack them.

Blood clotting

If blood vessels are damaged, a series (cascade) of enzyme-controlled reactions occurs to form a clot – preventing further blood loss and invasion by pathogens.

- Clotting depends on **clotting factors**, which are **plasma enzymes**.
- These are present in inactive forms in the **blood plasma**.
- They are named with roman numerals – Factor I, II ...V, VI ... up to X.
- There are two clotting pathways: one involves platelets, the other does not.

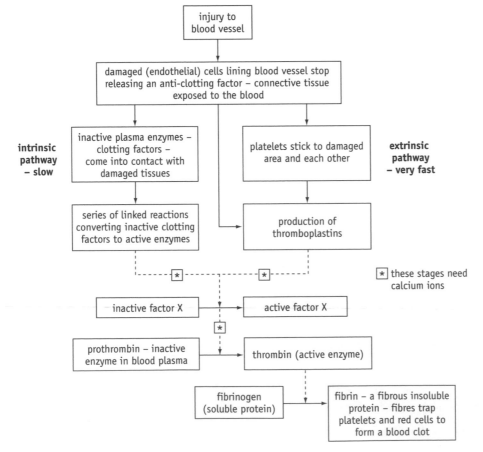

Formation of a blood clot

> Haemophilia is an inherited disease where the sufferers make non-functional factor VIII. If they get cut, they bleed for a long time. Treatment involves injecting the factor.

> A flow chart like this is a good way to remember facts, but you need to be able to explain what is going on – not just learn and reproduce the chart.

✓ *Quick check 1, 2*

Phagocytosis

Some lymphocytes are called **phagocytes**. These cells attack pathogens which have antibodies attached to antigens on their surface. (See next section about antigens and antibody production.)

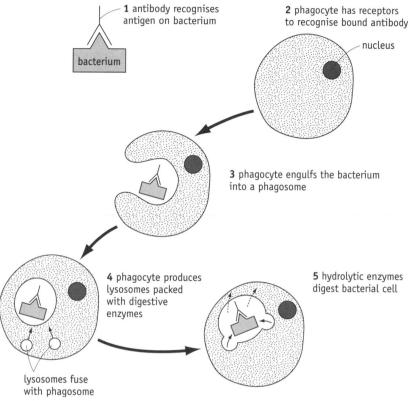

1 antibody recognises antigen on bacterium

bacterium

2 phagocyte has receptors to recognise bound antibody

nucleus

3 phagocyte engulfs the bacterium into a phagosome

4 phagocyte produces lysosomes packed with digestive enzymes

5 hydrolytic enzymes digest bacterial cell

lysosomes fuse with phagosome

Phagocytosis and destruction of pathogens by a white blood cell (phagocyte)

✓ *Quick check 3*

? *Quick check questions*

1 Explain the role of calcium ions in blood clotting.

2 Explain the role of thrombin in blood clotting.

3 Explain how phagocytes destroy pathogenic bacteria.

HB

Principles of immunology

The body defends itself against invading foreign organisms, cells and large organic molecules. This **immune response** involves **lymphocytes**.

Humoral and cellular immune responses

There are many types of lymphocyte. **B-lymphocytes** are responsible for **humoral responses** – the production and **release of antibodies** into blood plasma in response to an antigen. **T-lymphocytes** are responsible for **cellular responses** – they **bind to antigen-carrying cells** and destroy them, and/or activate a humoral response.

Antigen and antibody

An **antigen** (or immunogen):

- causes an immune response;

- is a protein, polysaccharide or glycoprotein;

- is usually on the cell wall or plasma membrane of cells.

An **antibody**:

- is a protein (immunoglobulin) produced by a B-lymphocyte;

- binds to a specific antigen because of the tertiary structure (3-dimensional shape) of the antibody.

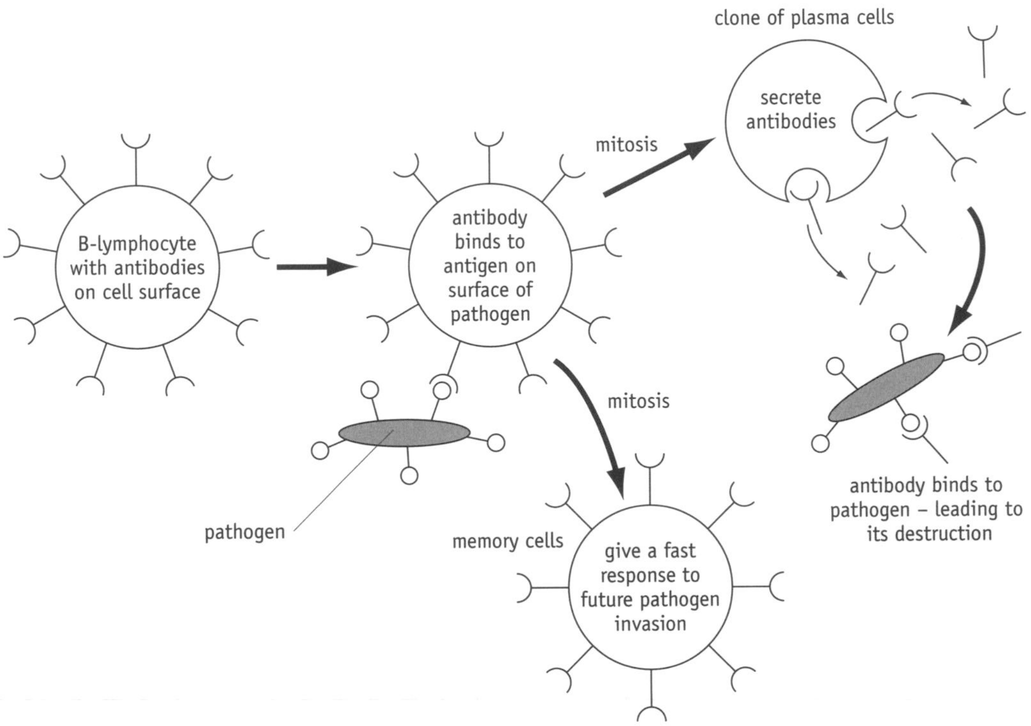

Binding of the antibody usually leads to the **destruction of the antigen** – and any cell attached to it, such as a **pathogen**.

Antibody production by a B-lymphocyte – plasma and memory cells

Humoral response by B-lymphocytes

- B-lymphocytes are one type of white blood cell.

- As each B-lymphocyte differentiates, it produces its own **specific antibody**, inserted in its **plasma membrane**.

If a B-lymphocyte meets its specific antigen:

- the antigen binds to a **receptor site** on the antibody in the membrane;

- the antigen fits the shape of the receptor.

✓ *Quick check 1*

As with any protein, the function of an antibody is linked to its tertiary structure – its 3-dimensional shape. You should look at questions with this in mind.

Binding makes the B-lymphocyte:

- **start secreting antibody** into the blood;

- **start dividing by mitosis**, producing a **clone** (clonal expansion) of very many genetically identical B-lymphocytes called **plasma cells**.

- These plasma cells secrete **lots of the same antibody** – leading to destruction of the antigen.

This is the **primary (1°) response. Memory cells** are formed from the clone.

A **secondary (2°) response** happens:

- if the **same antigen** enters the body again after some time – perhaps many years;

- immediate recognition and destruction occur – the person has **active immunity** – giving a **fast response** and **preventing harm by the antigen**.

> ▶ If a B-lymphocyte does not meet its antigen, it dies. The body does not produce a specific B-lymphocyte in response to invasion by an antigen. The body produces an almost infinite number of different B-lymphocytes, so that hopefully there will be one that binds to any antigen that invades.
>
> ✓ *Quick check 2*

Cellular response by T-lymphocytes

T-lymphocytes are responsible for **cell-mediated immunity**. They respond to microorganisms such as protozoa, fungi or larger bacteria and infected host cells. They also regulate the activity of B-lymphocytes.

- An antigen on the surface of a cell binds to a specific **antigen receptor** (which is not an antibody) on the plasma membrane of a T-cell.

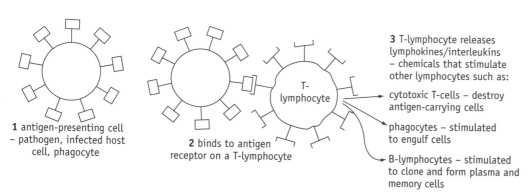

1 antigen-presenting cell – pathogen, infected host cell, phagocyte

2 binds to antigen receptor on a T-lymphocyte

T-lymphocyte

3 T-lymphocyte releases lymphokines/interleukins – chemicals that stimulate other lymphocytes such as:

cytotoxic T-cells – destroy antigen-carrying cells

phagocytes – stimulated to engulf cells

B-lymphocytes – stimulated to clone and form plasma and memory cells

T-lymphocyte response to antigen

> ▶ HIV destroys T-lymphocytes. Eventually this prevents an immune response to common opportunistic infections and the person dies – usually from a combination of infections.

- This leads to the reactions shown in the diagram.

> ✓ *Quick check 3*

? *Quick check questions*

1 Explain what happens when a B-lymphocyte meets its antigen.

2 Explain why the humoral response depends on the cellular response to antigen.

3 HIV destroys T-lymphocytes. Explain how HIV infection leads to AIDS and the inability to fight other infections.

HB

Passive and active immunity

Passive immunity

Passive immunity is a **short-lived** defence against an antigen (or antigen-carrying pathogen) by antibodies introduced into the body from outside.

Natural passive immunity

- **Babies** get **antibodies** from their mother across the placenta and via **lactation** (breast milk).
- The baby gets short-term defence against some common pathogens until its own immune system starts to work.

Artificial passive immunity

- By **injection** of antibody from an artificial source – produced in another animal.
- For example, anti-venom contains antibody against a snake venom antigen.

Passive immunity produces **no memory cells**. Any antibody not used soon disappears (within weeks), leaving the person vulnerable to the antigen again.

> Anti-venoms and anti-toxins are used when an antigen is likely to kill someone before their own immune system has time to react and produce its own antibody.

✓ *Quick check 1*

Active immunity

This gives long-term defence against an antigen (pathogen). It involves production of **memory cells** able to give a rapid secondary response to an antigen.

Active immunity is obtained by:

- natural exposure to the antigen (pathogen);
- **vaccination** – artificial introduction of a form of the antigen which causes an **immune response** but not serious illness.

Vaccination produces active **artificial** immunity – **immunisation**.

Different types of vaccine are shown in the table.

Type of vaccine	Example
Live non-virulent strain, from generations of selective subculturing in the laboratory	Rubella, tuberculosis
Killed virulent organism	Whooping cough
Chemically **modified toxin molecule** that is no longer toxic but still resembles the toxin antigenically	Diphtheria, tetanus
Isolated antigens, separated from the microorganism	Influenza
Mass-produced **antigen from genetically engineered bacteria**	Hepatitis B

✓ *Quick check 2*

The diagram shows how vaccines produce responses by the immune system.

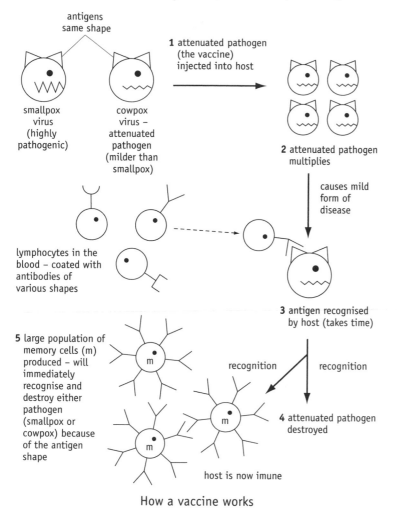

Since vaccines have to cause an immune response, there is always the chance that they might make someone ill. This chance is minimal compared with the risk from the disease itself.

How a vaccine works

✓ *Quick check 3*

? *Quick check questions*

1   Explain what is meant by passive immunity.

2   Describe three sources of vaccines.

3   Explain how vaccination leads to active immunity.

HB Heart disease

Heart disease and cancer are not passed on from person to person by pathogens. The risk of getting them is often affected by, and can be reduced by, modification of lifestyle.

The biological basis of heart disease

Heart disease is one of the main causes of death in the UK. One of the main forms is development of atheroma in **coronary arteries** that carry blood to heart muscle – leading to **coronary heart disease**.

Atheroma

This is a degeneration of the walls of the arteries (arteriosclerosis).

- Fatty deposits develop in artery walls – especially **cholesterol**.

- These enlarge into **plaques**.

- Some harden and cause narrowing of the artery.

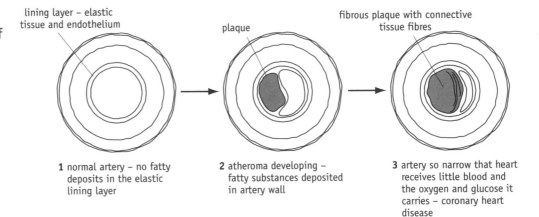

lining layer – elastic tissue and endothelium

plaque

fibrous plaque with connective tissue fibres

1 normal artery – no fatty deposits in the elastic lining layer

2 atheroma developing – fatty substances deposited in artery wall

3 artery so narrow that heart receives little blood and the oxygen and glucose it carries – coronary heart disease

Plaque formation in coronary artery

- No symptoms occur until the narrowing is enough to interfere with the circulation, especially during exercise.

- This causes **angina** – chest pain when exercising.

- It can lead to aneurysm, thrombosis or myocardial infarction.

- If the artery becomes completely blocked it leads to **heart attack**.

Aneurysm and thrombosis

The risk of developing these conditions is increased by atheroma formation.

- An **aneurysm** is a swelling of an artery (most commonly the aorta and the arteries supplying the brain).

- Atheroma causes **increased blood pressure** because the heart beats harder to force the blood through the narrower vessels.

- This extra pressure weakens the artery wall, causing it to distend or forcing blood through a fissure.

> If diagnosed before it bursts, an aortic aneurysm can be cut out and replaced with a piece of artificial artery – made of a nylon-like material.

✓ Quick check 1

> High blood pressure over long periods can cause splits in the lining of the artery – encouraging atheroma formation.

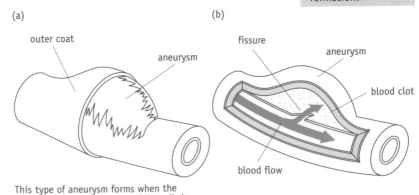

(a)

outer coat

aneurysm

(b)

fissure

aneurysm

blood clot

blood flow

This type of aneurysm forms when the tunica media, the artery's middle wall, is weakened; the strong force of the blood flow swells the wall of the artery

A common aneurysm

- A **thrombosis** is the formation of a **blood clot** in an artery, which blocks the flow of blood.
- Heart, brain and leg arteries are most commonly affected.
- The clot usually forms where **atheroma** has **damaged the arterial lining**.

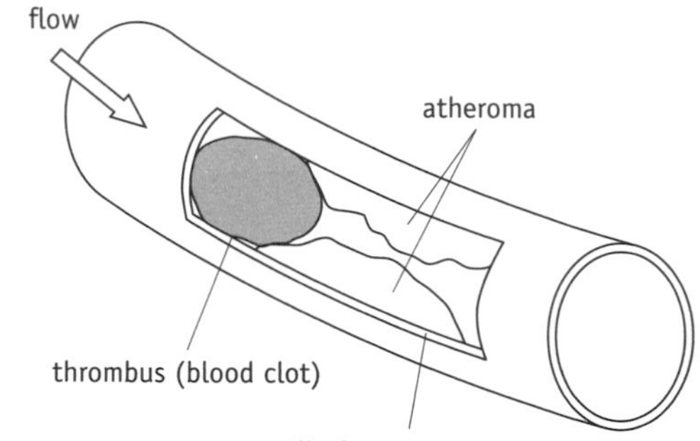

Myocardial infarction

This is a condition in which heart muscle dies when its blood supply is cut off.

- Atheroma forms in a **coronary artery** supplying part of the cardiac muscle.
- A coronary thrombosis occurs – a blood clot blocks a coronary artery.
- Heart muscle cells start to die due to **lack of oxygen**.
- Symptoms include severe chest pain that spreads to the arms and the jaw.
- It can be fatal, due to heart failure.

A thrombosis associated with atheroma

✓ *Quick check 2*

Risk factors associated with coronary heart disease

Coronary heart disease affects the coronary arteries.

The major factors linked to an increased risk of coronary heart disease are shown in the table.

Risk factor	Effect on heart disease
Diet rich in saturated fatty acids and/or cholesterol	• High blood cholesterol increases atheroma formation in coronary arteries
Smoking	• Nicotine causes constriction of blood vessels, leading to high blood pressure, damage to coronary arteries and more chance of atheroma formation. • Increased tendency of blood to clot and cause a thrombosis. • Free radicals accelerate atheroma formation.

There are other factors that increase the risk of developing coronary heart disease. These include; genetic factors, lack of exercise, obesity and diabetes.

✓ *Quick check 3*

? *Quick check questions*

1 Explain what atheroma is.
2 Explain why a coronary thrombosis leads to myocardial infarction.
3 Explain how two lifestyle factors affect the risk of developing heart disease.

HB

Cancer

The biological basis of cancer

Benign and malignant tumours

- A **tumour** is a mass of undifferentiated cells dividing out of control.
- Tumour **cells** are smaller than ordinary cells and have clear nuclei.
- They divide by **mitosis**.
- As a tumour grows it destroys the normal cells that surround it.
- **Benign** tumours do not spread to other tissues or organs – their cells stay together as a 'lump'.
- **Malignant** tumours undergo **metastasis** – they **invade other organs**.
- Cells break away from the primary tumour, are carried to other organs by the blood and lymphatic system and form secondary tumours (**metastases**).

> Cancer is not one disease; there are very many different types, with many causes. This means that one 'cure' for cancer is unlikely to be found.

> ✓ **Quick check 1**

Genes, the control of cell division and formation of tumours

- Normal cell growth is controlled by genes.
- Some genes code for the production of **growth factors** (cyclins) that stimulate cell division by mitosis.
- Other genes code for the cell surface membrane proteins that are **receptors** for the growth factors.
- A **mutation** of one of these genes can lead to uncontrolled cell division – cancer.
- The mutated gene is now an **oncogene** – a gene promoting cancer.

> Oncogenes can be inherited – giving a greater risk of developing cancer, for example some types of breast cancer.

a How a normal cell responds to growth factor   **b** How a mutated cell responds

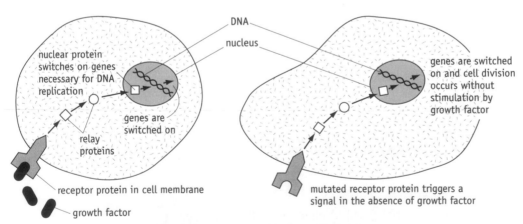

The effect of a mutation on cell division

> Some cancers are caused by viruses that bring oncogenes into cells. Other cancers – like bowel cancer – are linked to dietary factors.

- **Tumour suppressor genes** produce proteins that block the ability of growth factors to stimulate cell division – blocking growth of a tumour.
- A mutation of one of these genes can also lead to cancer.
- Cancer cells often have **abnormal proteins** in their cell surface membranes.
- These usually cause an **immune response** – the cells are destroyed by lymphocytes (white blood cells).
- Mutated cells that do *not* cause an immune response are the ones that cause cancer.

> *p53* is an important tumour suppressor gene. Mutated *p53* genes are found in over 50% of cancers.

> ✓ **Quick check 2**

Causes of mutations in genes controlling growth

Mutations of the DNA of regulator genes may occur by chance, but the rate is increased by:

◖ Mutations change the base sequence in DNA. This leads to changes in the tertiary structure of the protein produced – making it non-functional.

HB

- **high energy radiation**, including UV light (in sunlight) and X-rays;
- emissions from **radioactive substances**;
- **carcinogens** – chemicals found, for example, in **tar in tobacco smoke**.

Lung and skin cancer in the United Kingdom

In the UK 16% of cancers are of the lung and 9% of the skin (1998 figures).

- **Lung cancer** is associated with **cigarette smoking** – tar contains carcinogens (mutagens to the lungs).
- 96% of lung cancer victims who died in the 1980s were smokers.
- **Skin cancer** is linked to exposure to **UV light** – from the Sun or 'sunbeds'.
- The greater the exposure to UV light and the lighter the colour of your skin, the greater the risk of skin cancer.
- People in Australia are very much at risk as the natural filter of the ozone layer is thinnest near to the South Pole.

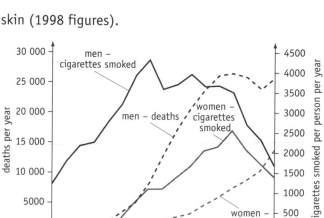

Graphs showing a correlation between smoking and lung cancer

✓ *Quick check 3, 4*

? *Quick check questions*

1 Explain the difference between a benign and a malignant cancer.

2 Explain why a cancer cell divides out of control.

3 **a** Describe the relationship between smoking and lung cancer in women.

  b Suggest why the number of lung cancer deaths per year was still rising amongst women in 1991, when the number of cigarettes smoked per person per year was falling.

4 Suggest why people living in Australia have a very high risk of developing skin cancer.

HB

Disease and diagnosis

DNA probes and diagnosis

Some **human diseases** are caused by a **specific gene** that is defective; for example, cystic fibrosis, sickle-cell anaemia and Huntington's chorea. For some of these diseases it is now possible to test to see if someone has the defective gene.

Use of restriction enzymes

- A sample of the person's DNA is **cut into small fragments** by **restriction** (endonuclease) **enzymes**.
- Each restriction enzyme is **specific** – it cuts DNA **wherever a specific base sequence occurs.**
- The gene being tested for will be in one of the fragments.

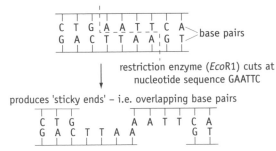

Action of a restriction enzyme

Electrophoresis and separation of DNA fragments

- DNA fragments are separated by **gel electrophoresis**.
- The mixture of fragments is placed in a well at the top of the gel.
- Smaller DNA fragments move further in the gel when an electrical potential is applied, as shown in the diagram.

Radioactive DNA probes

- A **nylon membrane** placed on the gel picks up a sample of the DNA fragments – still in order.
- DNA fragments are treated to separate their double helices into **single strands**.
- Radioactive **DNA probes** are then put on the nylon.
- Each probe is a single nucleotide strand with a **base sequence complementary to a particular defective gene**.
- The sugar-phosphate backbone of the probe is **radioactively labelled**.
- Each DNA probe binds to its gene by **complementary base pairing**.
- X-ray film is placed on the nylon and 'fogs' where radioactive DNA probes are present – giving a **black band**.
- If a black band appears, the person has the defective gene.

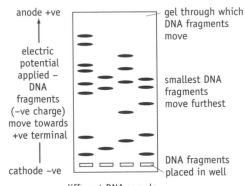

Gel electrophoresis

✓ **Quick check 1**

Enzymes and diagnosis

The enzyme associated with a particular metabolic pathway is only found in a certain part of the body, in a certain concentration range.

Some **diseases** cause **changes in the concentration of enzymes**, or lead to them being **found in abnormal places**. This can be used as a test for the disease.

Pancreatitis

The pancreas produces **digestive enzymes**, including **trypsin** and an **amylase**.

- These are released along the pancreatic duct into the small intestine.

- In some people, juices from the small intestine get forced back along the pancreatic duct.

- The digestive enzymes attack the pancreas, causing **inflammation**.

- This leads to **leakage of trypsin and amylase into the blood**.

- They are not active in the blood – amylase because it has no substrate in the blood plasma, trypsin because of inhibitors.

- A **blood sample** can be **tested for amylase** activity by adding the substrate (glycogen) and testing for the product (glucose).

- Pancreatitis can also lead to **lower enzyme concentrations in the intestine** – resulting in slower digestion.

✓ *Quick check 2*

Analytical reagents

▶ Make sure you know the enzyme, its substrate and the product at each stage of the test for glucose – they have to be in the right order and place!

- Enzymes are very **specific** – they only bind to their substrate.

- This means they can be used to **identify a specific substance** in a sample.

- Enzymes have **high sensitivity** – they react with **low concentrations of substrate**, to produce measurable amounts of product(s).

- An example of an enzyme used as an analytical reagent is **glucose oxidase** (see diagram).

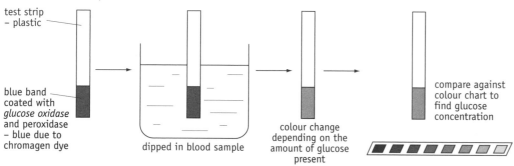

test strip – plastic

blue band coated with *glucose oxidase* and peroxidase – blue due to chromagen dye

dipped in blood sample

colour change depending on the amount of glucose present

compare against colour chart to find glucose concentration

Glucose + water + oxygen $\xrightarrow{\text{glucose oxidase}}$ gluconic acid + hydrogen peroxide

Blue chromagen dye + hydrogen peroxide $\xrightarrow{\text{peroxidase}}$ green to brown chromagen dye + water

Testing blood plasma for glucose

✓ *Quick check 3*

? *Quick check questions*

1 Explain how a DNA probe could be used to find out if someone carried the specific gene for a genetic disease.

2 Explain how the distribution of enzymes can be used to diagnose pancreatitis.

3 **a** Describe how you would use named enzymes to test for glucose.

   b You are given a blood sample from someone with suspected pancreatitis. Suggest how you might test this diagnosis.

HB

Drugs in the control and treatment of disease

Drugs are chemical substances used in medicine. Some act on symptoms of disease. Some kill pathogens, or slow down their growth. Others target substances or cells involved in disease, allowing them to be destroyed.

Beta blockers

These drugs **treat hypertension** (high blood pressure) – when someone has higher blood pressure than normal, often linked to stress, excessive drinking, obesity or smoking.

- **Heart muscle cells** have beta **receptor molecules** in their plasma membranes.

- When the body is under stress, **noradrenaline** is released.

- This **binds to beta receptors**, stimulating contraction of the heart muscle cells.

- This leads to higher blood pressure.

- **Beta blockers** bind to the beta receptor molecules instead of noradrenaline.

a without beta blocker

noradrenaline

beta receptor

stress → release of noradrenaline

cardiac (heart) muscle cell membrane

hypertension

noradrenaline in beta receptor stimulates heart muscle to contract

b with beta blocker

noradrenaline

beta blocker

stress → release of noradrenaline

beta blocker is correct shape to block the receptor but does not stimulate muscle to contract

no stimulation

no hypertension

Competition between beta blocker and noradrenaline for binding to receptor molecules

> Part of the beta blocker's shape allows it to fit into the receptor site, rather like a competitive inhibitor of an enzyme.

✓ *Quick check 1*

Antibiotics

Antibiotics are produced naturally by fungi, to kill or inhibit the growth of bacteria that compete with them for food. Medicinal antibiotics treat bacterial disease, leaving human cells unharmed.

Bacteriostatic antibiotics

- These **stop growth** and reproduction of bacterial cells.

- Some bind to bacterial ribosomes, preventing protein synthesis and thus growth, e.g. streptomycin. (Remember that bacterial ribosomes are different from those in human cells.)

- Others prevent DNA replication and transcription, e.g. ciproflaxin.

> Preventing transcription stops the production of enzymes needed for the metabolism of bacterial cells to work.

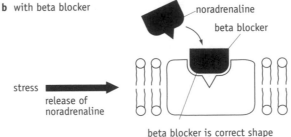

streptomycin binds specifically to bacterial ribosomes, preventing protein synthesis and thus growth

ciproflaxin prevents DNA replication and transcription of mRNA

penicillin prevents cell wall synthesis

How antibiotics work

Bactericidal antibiotics

- These antibiotics **kill** bacterial cells.
- Some inhibit enzymes needed to make bacterial cell walls, e.g. penicillin.
- The bacterial cells grow a weakened cell wall.
- Lysis (bursting of the cell) happens as the cells take up water by osmosis and swell without strong cell walls to prevent expansion.

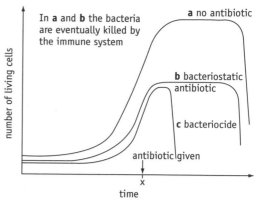

HB

The effect of adding bacteriostatic and bacteriocidal antibiotics on the growth of bacterial populations

✓ *Quick check 2*

Monoclonal antibodies

Monoclonal antibodies are **identical antibodies** produced by a **clone of a single B-lymphocyte**. They are **highly specific** to one antigen, and can bind to and **target a specific substance** (antigen), **or cells** carrying their antigen.

- **Pregnancy tests** use monoclonal antibody that binds to the hormone HCG (human chorionic gonadotrophin) found in the urine of pregnant women.
- A **'magic bullet'** is an antibody that binds to an antigen on cancer cells but not normal cells.
- A **cytotoxic drug** (cell poison) attached to the antibody kills the cancer cell.
- For **diagnosis** – antibody that binds to antigen on a cancer cell can have a marker attached to it, so cancer cells can be identified if they are present.
- The marker can be a radioactive element, or a fluorescent chemical visible under UV light.

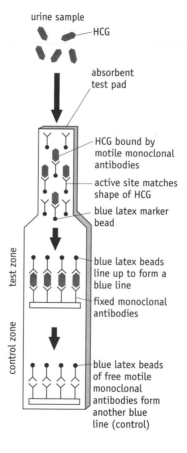

1 HCG present in urine sample of pregnant women

2 urine absorbed onto test pad activates the monoclonal antibodies, which become motile

3 monoclonal antibodies fixed in the test zone bind with the motile HCG/antibody complexes, resulting in a thin blue line in the test zone (this is a positive test)

4 any free motile monoclonal antibodies are 'mopped up' in the control zone resulting in a second blue line (proof the test is working) – the only blue line if the woman is not pregnant

Pregnancy testing

✓ *Quick check 3*

? ## Quick check questions

1 Explain how a beta blocker works.
2 A bacteriostatic antibiotic only prevents the growth of bacteria in the body. Suggest how this helps the patient to avoid serious illness.
3 **a** Explain what is meant by a monoclonal antibody.
   b Describe and explain one use of a monoclonal antibody.

'Magic bullets' are very promising for the future but have had limited success so far. The problem is in getting an antibody that binds to an antigen found only on cancerous cells and not normal cells of the patient.

Module 3: end-of-module questions

1   Samples of DNA were obtained from a child and his parents. The child was tested
    for the presence of a specific gene associated with a human disease, using a
    radioactive DNA probe. The diagram shows the result of the analysis of their DNA.

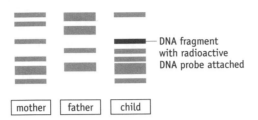

    DNA fragment
    with radioactive
    DNA probe attached

    | mother | father | child |

   a   Each band represents a fragment of DNA. Explain how these
         fragments were:

      i made from the DNA samples;   [2]

      ii separated by gel electrophoresis.   [3]

   b   Explain how the radioactive DNA probe binds to a specific gene.   [2]

   c   Suggest why the pattern of banding for the child is not the same
         as either parent but has similarities to both.   [2]

   d   Explain which parent the child inherited the gene from.   [1]

2   The diagram shows the structure of HIV.

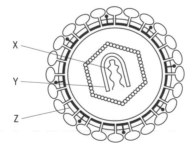

    Name parts X, Y and Z and explain their function in replication of HIV.   [6]

3   The flow chart shows some of the reactions involved in the clotting of blood.

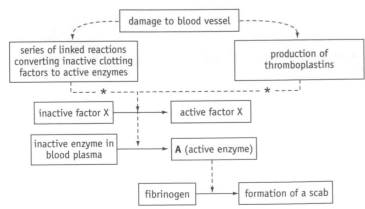

a Explain how damage to a blood vessel leads to the production of active clotting factors and thromboplastin. [3]

b Two stages are marked with * to show that calcium ions are needed. Put a **C** on the chart to show another reaction requiring calcium. [1]

c i Name substance **A**. [1]

ii Explain what happens when fibrinogen is acted on by enzyme **A**. [3]

d A child was diagnosed with an inherited condition which meant that he produced a non-functional factor X. Suggest the effect that this would have on the child. [3]

4 For each of the following, describe two modifications of their life cycle related to their parasitic way of life.

a *Plasmodium* [4]

b *Schistosoma* [4]

5 **a** i Explain what is meant by atheroma. [2]

ii Explain how two life style factors affect atheroma formation. [4]

b i Explain what is meant by an aneurysm. [1]

ii Explain the link between atheroma and the increased risk of aneurysm and thrombosis. [4]

c Suggest why a coronary thrombosis is so life-threatening. [2]

6 A tumour is a mass of cells that are dividing out of control. Some tumours are benign and others are malignant.

a Explain why tumour cells divide out of control. [5]

b Describe the difference between a benign and a malignant tumour. [3]

7 The graph shows data about the number of cigarettes smoked per year by male smokers and the number of cases per year of lung cancer in men.

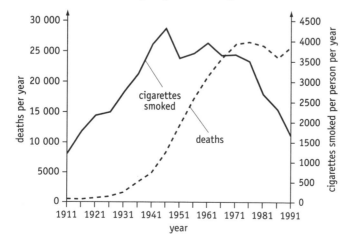

a Describe the relationship between the number of cigarettes smoked per year and the number of cases of lung cancer per year. [3]

b Explain the relationship between smoking and lung cancer. [4]

Appendix: Exam Tips

Lots of marks are lost by not answering questions as they are set, or not knowing material in the syllabus. You must know the information, terms and examples that are included in the syllabus – the exam board only allows questions that can be answered using syllabus material. Other information will not harm you, but will not be necessary to get a good mark!

Describe

'Describe' means put information into words. The information is usually **given** to you in a table, graph or diagram.

Example: the graph shows the rate of reaction of an enzyme with different concentrations of substrate and how the rate is affected by the addition of a particular concentration of an inhibitor.

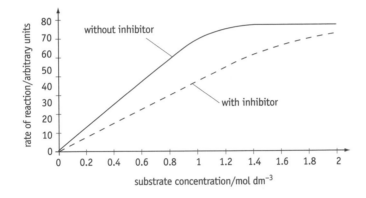

Question

Describe the effect of each of the following on the rate of reaction of the enzyme:

a the concentration of substrate;

b the inhibitor.

Answers

a Between substrate concentrations of 0 and 1 mol dm^{-3}, the rate is directly proportional to the concentration of substrate. The rate reaches a maximum at a substrate concentration of about 1.4 mol dm^{-3}.

b The inhibitor reduces the rate of action of the enzyme at lower concentrations of substrate. Above substrate concentrations of 1.2 mol dm^{-3}, the rate with the inhibitor gets closer and closer to the rate without the inhibitor.

These answers **describe** what you can see on the graph – in reasonable detail.

Explain

'Explain' means that you should 'know' the answer from syllabus material that you have been taught. The following question uses the same graph and information as for 'Describe'.

Question

Is the inhibitor competitive or non-competitive? **Explain** your answer.

Answer

The inhibitor is competitive, because its effect is overcome by increasing the concentration of substrate.

A non-competitive inhibitor would inhibit the enzyme at any concentration of substrate.

Suggest

'Suggest' means that you are unlikely to have seen the material in the question but you should have been taught things from the syllabus that will allow you to answer. The following question applies to an enzyme and its inhibitor.

Question

Two substances, **X** and **Y**, were investigated as possible rat poisons. Both inhibit the same enzyme in an important biological process. The diagrams show the structure of the enzyme, its substrate and the inhibitors. Use the information in the diagrams to suggest which inhibitor would be the best poison.

Answer

Inhibitor Y would be best, because it is non-competitive. It binds to a site other than the active site.

Its effect cannot be overcome by more substrate (unlike inhibitor Y) and so the metabolic pathway is blocked.

OR

Inhibitor X would be best, because it is competitive. It binds to the active site. This reduces/stops the substrate being turned into product (and stops the metabolic pathway).

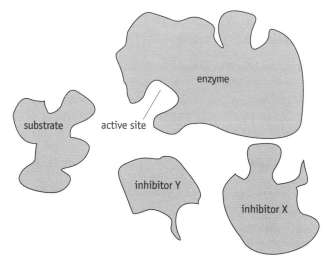

Rat poison

It is not unusual to have alternative answers to this sort of question. You are not expected to have learnt this material – it isn't specifically given in the syllabus. Either answer is a reasonable interpretation of the information.

Answers to quick check questions

Module 1: Molecules, Cells and Systems B   HB

Cells and microscopy

1

	Prokaryotic	Eukaryotic
	DNA is circular, not in a nucleus	DNA is linear, within a nucleus
	Diameter of cell 0.5–10 μm	Diameter of cell 10–100 μm
	Smaller, lighter, 70S ribosomes	Larger, heavier, 80S ribosomes
	No mitochondria present	Mitochondria present
	No golgi body	Golgi body present
	Flagella (when present) simple, lacking microtubules	Flagella (when present) have '9+2' arrangement of microtubules

2 Advantages – improved resolution / higher *useful* magnification. Disadvantages – specimens are dead / techniques can produce artefacts / expensive.

Cell ultrastructure – organelles

1 Both contain DNA / ribosomes / possess an envelope of two membranes.

2 (a) Protein synthesis (b) Production and transport of lipids (c) Produces glycoproteins / secretes enzymes / produces lysosomes / forms cell walls.

Cell fractionation

1 a Prevents action of enzymes that may break down cellular components.

   b Prevents osmotic movement of water which may cause organelles to burst or shrivel.

2 Nuclei

The plasma membrane

1 Lipid-soluble molecules can pass directly through/are soluble in the phospholipid bilayer.

2 a Transport ions and polar molecules across the membrane.

   b Receptors allow specific hormones to attach and stimulate cells.

Transport across membranes

1 Zero

2 Water moves by osmosis from a higher water potential in the soil into the root hair, which is at a lower water potential.

3 Diffusion is a passive process, active transport requires energy. Diffusion occurs along a concentration gradient, active transport occurs against a concentration gradient.

4 Lower the temperature / reduce the amount of oxygen available / add a metabolic inhibitor.

Carbohydrates – mono- and disaccharides

1 Glycosidic bond.

2 By a condensation reaction involving the loss of water and the formation of a glycosidic bond.

3 Heat with Benedict's solution; a brick-red colour is positive.

Carbohydrates – polysaccharides

1 Glucose molecules are joined together to form long straight chains. Hydrogen bonds link chains together to form microfibrils. Microfibrils are grouped together to form macrofibrils which provide rigidity.

2 Starch is insoluble and therefore osmotically inactive. Helical shape of starch molecule forms a compact store. Contains a large number of glucose molecules providing abundant respiratory substrate.

3 Liver or muscle.

Proteins

1 Nitrogen

2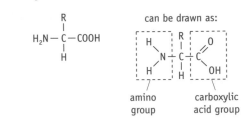

3 The sequence of amino acids in a polypeptide chain.

4 Denaturation occurs as hydrogen and ionic bonds are broken, leading to irreversible alteration of tertiary structure.

Lipids

1 A fatty acid that does not possess any double bonds.

2 White emulsion.

3 A triglyceride molecule consists of a glycerol molecule with three fatty acids. In a phospholipid one of the fatty acids is replaced by a phosphate group.

Chromatography

1 R_f value $= \dfrac{\text{distance moved by compound from origin}}{\text{distance moved by solvent from origin}}$

2 Locating agents can be used to stain colourless spots.

3 Use 2-way chromatography. Spot the mixture at the origin and run a chromatogram with a suitable solvent. Record R_f values, turn the chromatogram 90° and run a second chromatogram using a different solvent. Record R_f values.

Enzymes

1 Enzymes possess a specific tertiary structure that determines the shape of the active site. Lipids have the complementary shape to bind at the active site of lipases, but carbohydrates have a different shape and cannot bind.

2 Enzyme concentration / substrate concentration / temperature / pH.

3 Denaturation occurs as hydrogen and ionic bonds are broken, leading to irreversible alteration of the tertiary structure. The shape of the active site is altered and the substrate can no longer bind.

4 Add more substrate – if a competitive inhibitor is present the rate of reaction will increase; if a non-competitive inhibitor is present no change will occur.

Tissues and organs

1 a Cells of same type, working together for a function.

   b A collection of tissues working together for a function.

2 Epithelium is, thin (one cell thick) – short diffusion pathway; collectively it has large surface area.

3 Biconcave disc – large surface area for exchange; filled with haemoglobin – binds to/carries oxygen; no nucleus – more room for haemoglobin.

4 Flat in cross-section, so no cell far from exterior; short diffusion pathway for gas exchange; flattened shape – large surface area.

Blood vessels

1 Twice as thick.

2 Artery thick wall and narrow lumen – blood flows rapidly under high pressure; vein large lumen, thin wall – blood flows slowly at low pressure; same net flow rate past a given point per minute.

3 Artery near to surface of body; feel bulge in artery wall; as each surge of blood flows past, corresponding to one heartbeat.

Circulation

1 We are large organisms with a small surface area to volume ratio; so we need special internal exchange surfaces; need a mass flow system to carry substances between exchange surfaces and rest of body – blood system.

2 Hepatic portal vein to liver, hepatic vein to vena cava to right side of heart, pulmonary artery to lungs, lung capillaries, pulmonary vein to left side of heart, aorta, then renal artery.

3 Venule end of capillary – low hydrostatic pressure, blood proteins give low water potential for blood plasma – lower than tissue fluid – so water reabsorbed into plasma by osmosis.

Lung function

1 Thin – short diffusion pathway; large surface area relative to the volume of the organism; steep diffusion gradient – maintained by transport/blood system.

2 Intercostal muscles and diaphragm contract; moves ribcage up and out and flattens diaphragm; increases volume and lowers pressure in thorax/lungs; to below atmospheric pressure, so air forced in along pressure gradient.

3 Inspiratory centre causes inspiration; lung inflation stimulates stretch receptors; nerve impulses to medulla; inhibit the inspiratory centre, leading to expiration; stretch receptors no longer stimulated; inhibition of inspiratory centre lifted.

Heart structure and function

1 Atrioventricular valves prevent backflow from ventricles to atria; semilunar valves prevent backflow from arteries into ventricles.

2 Sinoatrial node acts first, so atria contract before ventricles – so blood enters relaxed ventricles; atrioventricular node reacts when ventricles are full.

3 0.8 seconds for one beat; $60 \div 0.8 = 75$ beats per minute.

4 Atrioventricular valve only opens when pressure inside ventricle is below that in the atrium.

Effects of exercise

1 a The volume of blood pumped by each ventricle per minute.

   b The volume of air breathed in per minute.

2 Rate of beating and stroke volume increase – increasing cardiac output.

3 More muscle contraction needs more energy from more respiration; more carbon dioxide produced – lowers blood pH; detected by chemoreceptors in aortic body, carotid body and medulla; leads to increase in breathing rate and depth of each breath.

4 (a) Gut –25%; skeletal muscle 2567%. (b) Blood supply to gut reduced, so more blood can be sent to skeletal muscles; increased supply to muscles to provide more oxygen and glucose for respiration. (c) After a meal, more blood to gut to carry absorbed food to liver; sleepy feelings reduce probability of vigorous exercise, which would cause blood to flow to skeletal muscles.

Module 2: Making Use of Biology

Mitosis

1   For growth and asexual reproduction.

2   Represents the total number of chromosomes in a normal body cell. In humans this is 46, i.e. 23 homologous pairs.

3   (**a**) Anaphase (**b**) Prophase (**c**) Interphase

Mitosis and the cell cycle

1   Interphase, mitosis and cytokinesis.

2   DNA replication.

3   Acetic (ethanoic) orcein.

Meiosis

1   4

2   A reproductive cell, e.g. a sperm or ovum.

3   In dipolid organisms, meiosis ensures that haploid gametes are produced. When the haploid gametes fuse at fertilisation to form a zygote the diploid number is restored, ensuring that each generation possesses a constant number of chromosomes.

Structure of nucleic acids

1   Pentose, phosphoric acid and an organic base.

2   Adenine, thymine, cytosine and guanine.

3   By specific hydrogen bonds.

4   **a**   RNA is a single strand, DNA is a double strand. / RNA contains ribose, DNA contains deoxyribose. / RNA has the base uracil, DNA has thymine.

   b   tRNA is 'clover leaf' in shape. / tRNA has standard length, mRNA is variable. / tRNA has an amino acid binding site. / tRNA has anticodons, mRNA has codons. / tRNA has hydrogen bonds between base pairs.

DNA – replication and the genetic code

1   When DNA replicates, each newly formed DNA molecule contains one of the original polynucleotide strands.

2   DNA nucleotides are joined together by the enzyme DNA polymerase to form complementary strands to the original DNA strands.

3   A different form of a particular gene.

4   The sequence of DNA nucleotide bases coding for the production of a specific polypeptide.

5   Some amino acids are coded for by more than one codon.

DNA and protein synthesis

1   A A C G C U G C A C C G U

2   DNA replication involves two template strands, transcription involves one. / DNA replication uses DNA polymerase, transcription uses RNA polymerase. / DNA replication uses DNA nucleotides, transcription uses RNA nucleotides.

3   Protein/polypeptide.

4   Carry specific amino acids to the ribosomes; have anticodons that bind to codons on mRNA.

Recombinant DNA

1   **a**   Cut DNA at specific base sequences to obtain a specific gene or to cut a plasmid vector molecule.

   b   The plasmid DNA and the 'foreign' DNA (gene) are joined together using a ligase enzyme that fits together the overlapping base pairs ('sticky ends').

2   Plasmids are small circular sections of DNA found in some bacteria and often used as vectors in genetic engineering.

3   Genetic markers in genetically engineered bacteria enable them to survive exposure to specific chemicals whereas non-recombinant bacteria are destroyed.

4   Transfer of foreign genes to non-target organisms. / Unknown ecological and evolutionary consequences. / Development of new resistant species, e.g. antibiotic-resistant bacteria. / Accidental transfer of unwanted genes by the vector, e.g. virus.

Isolation of enzymes

1   Bacteria and fungi secrete extracellular enzymes; easier to isolate than intracellular; secreted in large amounts into growth medium; often secreted on their own; more stable; purification/downstream processing easier.

2   e.g. Pectinase: filter growth medium to remove cells; evaporate to concentrate enzyme; precipitate pectinase as a solid; dry and grind to powder.

3   **a**   Makes the fungus produce more amylase.

   b   To avoid contamination with unwanted microorganisms; producing unwanted/harmful substances that contaminate the enzyme; increasing costs of purification; reducing yield by harming/competing with fungus producing the enzyme.

Application of enzymes in biotechnology

1   e.g. Glucose oxidase to test for glucose: glucose in urine/blood; glucose reacts to form gluconic acid and hydrogen peroxide; peroxidase reacts hydrogen peroxide and blue chromagen dye; colour change green to brown depending on concentration of glucose; colour compared against calibrated colour chart to find concentration.

2   e.g. Fructose production: amylase acting on starch at 100°C to give short chains of glucose; amyloglucosidase acts on chains at 55°C to give glucose; glucose isomerase acts on glucose at 60°C to give fructose; high temperatures give rapid reaction and production rates; fructose very much sweeter than glucose; valuable in diet foods, where less needs to be added.

3 Can be used in continuous fermenters; for long periods of time, since not removed with products; downstream processing easier and cheaper, because no contamination of product with enzyme; enzymes more heat and pH stable, so less need for very fine control of fermenter environment and can run at higher temperature.

4 Relative to the value of the product – cost of the enzyme; how easy is it to maintain optimum pH and temperature; how easy is purification of product/ downstream processing; can the enzyme be immobilised?

Immunology and blood groups

1 a Causes an immune response; protein/ polysaccharide or glycoprotein.

b A protein; produced by a B-lymphocyte; binds to and destroys the antigen.

2 Antibody is a protein; each has a unique tertiary structure/3-D shape; shape fits a specific antigen.

3 A B-lymphocyte has its antibody on its plasma membrane; if the specific antigen binds to the antibody, the cell starts to secrete antibody; divides (mitosis) to form a clone of plasma cells; these all secrete the antibody; memory cells also formed.

4 Injected antigen causes an immune response; antibody to antigen produced from plasma cells; memory cells formed; produce a rapid immune response to a later infection by the actual pathogen.

5 Anti-A antibody in type O person will agglutinate donated A blood, blocking blood vessels.

Genetic fingerprints and the polymerase reaction

1 Short repeated sequences of bases of non-functional DNA; VNTRs; position of these VNTRs is unique for each person.

2 Restriction enzymes to cut DNA into fragments; each enzyme cuts DNA wherever a specific base sequence occurs; using same restriction enzymes, each person's DNA cut into unique set of fragments.

3 Nylon blot, then treat DNA to give single strands. Add radioactive DNA probes – each specific to particular VNTR. Probes attach by complementary base pairing. Place X-ray film on top – fogging and unique set of bands.

4 a Polymerase chain reaction; separate DNA strands by heating at 95°C; cool to 60–70°C in the presence of nucleotides and DNA polymerase; new complementary strands formed; repeat heating and cooling cycle to double DNA each cycle.

b Each person has VNTRs; position of these is unique to each individual; cutting DNA samples with the same restriction enzymes; produces cuts in DNA in different places for each person; but the same position each time for an individual.

Adaptations of cereals

1 Structural – aerenchyma; air spaces in stem and roots; allows oxygen to diffuse to submerged roots for respiration. Physiological – root cells tolerant to ethanol; submerged roots can respire anaerobically without poisoning themselves with ethanol.

2 Different enzyme for 'fixing' carbon dioxide; high affinity for carbon dioxide; can fix carbon dioxide when concentration low in the leaf; at high temperatures stomata close more to save water – reducing carbon dioxide concentration in the leaf.

3 Cuticle reduces water loss by evaporation from the surface of the leaves – saving water; stomata where most water vapour is lost by diffusion; fewer stomata means less water loss; stomata sunk in pits to reduce air movement past them; so higher concentration of water vapour collects just outside stoma – reducing water potential gradient for diffusion of water.

The abiotic environment

1 When light intensity is not limiting; and when there is more than enough carbon dioxide.

2 To save money on fuel: on sunny days light not a limiting factor but it is on dull days, when a higher temperature will have no effect.

3 To replace mineral ions removed with crops; to maintain high yields; mainly nitrate, phosphate and potassium ions.

4 a Nitrate increased wheat yield; by about 3 times.

b Much greater yield than nitrate alone; increasing yields from 1975 to 1991.

5 Extra nitrate and phosphate ions cause eutrophication; cause a rapid growth of algae, which smother and kill other water plants; this damages food chains and webs relying on those plants; algae grow and then die in large numbers, leading to a rapid growth in populations of decomposing bacteria; bacteria use a lot of oxygen from the water for respiration – lowering oxygen in water for other organisms; many fish and invertebrates die of suffocation.

Pest control

1 Interspecific competition; for light, water, mineral ions, space or carbon dioxide; very important at seedling stage when root and shoot systems established; many weed seeds germinate earlier, or seedlings grow faster, than crop; they out-compete crop seedlings for some vital factor.

2 a By eating or damaging the part of the plant used by humans; examples, eating grain/eating or spoiling fruit.

b By reducing photosynthetic tissues, so there is less energy to put into fruit/grain, etc.

3 Biological control kills only certain pests, whereas chemical pesticide kills many different pests. / Only controls the number of pests, whereas chemical control can remove them completely (for a time). / Only reduces to below economically damaging levels, but this still reduces profits.

4 a Using several different methods of pest control together: chemical pesticides, pest-resistant crops, farming techniques and cultivation techniques together.

b repeated use of one method can allow a pest to adapt, or certain pests might not be controlled; using different methods together makes it more likely that more pests will be controlled/a particular pest will be controlled.

Environmental issues of pest control

1 a To simplify food webs involving crops/animals useful to man; to reduce competition by other organisms for our crops/animals; to increase yields and profits.

  b Many 'pests' are food for other organisms. / Weeds are the producers in food chains/webs. / Harmless organisms may also be killed. / Natural predators of pests can be killed as well. / Pollinating insects can be killed.

  c Slugs are a main food of frogs, so fewer slugs means fewer frogs; fewer frogs means less predation on slugs in future, so an increase in slug numbers.

2 An increase in the concentrations of pesticide in organisms as you go along a food chain; some pesticides not broken down in organisms or excreted, so accumulate; when organism is eaten, the pesticide is passed on and accumulates in the next organism in the food chain.

3 $(55 - 10)/55 \times 100 = 450\%$ more in hawk.

Reproduction and its hormonal control

1 One ovarian follicle ripens; primary oocyte inside finishes first meiotic division to form secondary oocyte; large, fluid-filled Graafian follicle forms; this ruptures, releasing the secondary oocyte/egg, which enters the fallopian tube.

2 Endometrium breaks down at menstruation; then starts to grow thicker; becomes more glandular; development of blood vessels/arteries and veins; greatest thickness about ten days after ovulation; if no pregnancy, then it starts to break down again.

3 a Peak of oestrogen production; causes rapid surge in LH (and FSH), which causes ovulation.

  b FSH secretion stimulates oestrogen secretion by the ovary; rise in oestrogen inhibits FSH release (from the pituitary gland); secretion of FSH leads to a reduction in its own release – negative feedback.

Manipulation and control of reproduction

1 a Changes in behaviour: hyperactivity; attempting to mount other cows; allowing other cows/bull to mount.

  b Selective breeding: inject cow with prostaglandin to bring into oestrus; inject with FSH, making many follicles ripen; artificially inseminate with sperm from superior bull; many embryos implant in the uterus; wash these out of uterus and transplant into healthy cows; many calves from valuable cow in short time.

2 Introduce a castrated ram to the flock; this visual stimulus causes the release of FSH and LH from the pituitary glands of the ewes; this brings the ewes into oestrus.

3 Blocked fallopian tubes, so no fertilisation possible; use in vitro fertilisation; inject woman with FSH, to make many 'eggs' ripen; remove 'eggs' surgically from ripe follicles; fertilise with man's sperm in vitro; transplant 2/3 embryos into the woman's uterus.

Module 3: Pathogens and Disease `HB`

Microorganisms and disease

1 Two named examples from the table on page 72.

2 To prove that a disease is caused by a microorganism, rather than some other cause – genetic, environmental.

3 Lag phase – appropriate enzymes being produced for new environment; log phase – no limiting factors, so exponential growth; stationary phase – population constant due to one or more limiting factors; decline/death phase – more organisms dying than are produced.

Parasites and parasitism (1)

1 Three examples from the table on page 74.

2 HIV has RNA as its genetic material; inside host cell reverse transcriptase makes DNA complementary to the HIV RNA; this DNA can then insert into the host cell's DNA and be replicated.

Parasites and parasitism (2)

1 *Plasmodium* reproduces inside host cells – liver and red cells; at intervals host cells burst and die – releasing parasites to infect more cells; death of host cells leads to symptoms of high fever and weakness.

2 Snail acts as an intermediate host in which the parasite reproduces; lots of infective stage produced that can enter through skin of human host.

Clotting and phagocytosis

1 Needed to produce active factor X; needed with factor X to convert prothrombin to thrombin.

2 This converts fibrinogen into fibrin, which forms the basis of a blood clot.

3 Antibody produced by the host attaches to antigens on the surface of bacteria; phagocytes have receptors that recognise bound antibody; they engulf bacterium and attached antibody; phagosome formed into which digestive enzymes are released; bacterium digested.

Principles of immunology

1 Antibody on plasma membrane of B-lymphocyte binds to its specific antigen; B-lymphocyte starts to secrete antibody and clones; these produce plasma cells – making the same antibody; clones produce memory cells that stay in the body for years and give a rapid response to subsequent invasion by the same antigen.

2 Cellular response by T-lymphocytes: when they bind to an antigen, they release lymphokines/interleukins that stimulate B-lymphocyte binding to the same antigen to clone; this gives plasma cells and the humoral immune response.

3 Without T-lymphocytes, no lymphokines/interleukins are produced; B-lymphocytes cannot clone to form plasma and memory cells; antibody is not produced to bind to antigen on pathogens; the pathogen can replicate itself and/or produce toxins with nothing to stop it, so eventually the person dies.

Passive and active immunity

1 Defence against an antigen/pathogen by antibody introduced into the body; natural passive immunity, e.g. from mother to baby across the placenta/in breast milk; artificial passive immunity, e.g. by injection of antibody such as anti-venom.

2 Three examples from the table on page 82.

3 Vaccine contains antigen; causes reaction by T- and B-lymphocytes that can bind to that specific antigen; B-lymphocyte clones to give plasma and memory cells; memory cells remain for years and allow body to produce antibody very rapidly if there is an invasion by the real pathogen; this rapid reaction is active – involves antibody formation.

Heart disease

1 Fatty deposits develop in the wall of an artery; enlarge to form plaque; can harden and cause narrowing of the lumen of the artery; this restricts blood flow.

2 Plaque in a coronary artery can cause a clotting reaction in the blood; a blood clot – a thrombosis – forms and blocks the artery where the plaque narrows the lumen; this stops the blood supply to an area of heart muscle; the muscle starts to die due to lack of oxygen (and glucose) – myocardial infarction.

3 Diet and smoking – with detail from the table on page 85.

Cancer

1 Benign tumour grows but stays as a single growth; malignant tumour – cells detach, travel to other organs and form secondary tumours.

2 Normal cell growth and division are controlled by genes for growth factors and receptors for factors; mutations of these genes can lead to uncontrolled cell division, forming a tumour; mutated genes are oncogenes; mutated tumour suppressor genes are often a cause.

3 a Rise in smoking from 1911 to 1940s followed by later rise in lung cancer – two lines roughly parallel; decline in smoking from 1940s followed by levelling off of incidence of lung cancer.

  b Women with lung cancer were those who smoked years earlier – it takes years for smoking to cause lung cancer.

4 Mainly European immigrants in Australia who have fair skin; Australia (near to equator) has strong sunshine with high intensity of UV light; this is mutagenic and increases risk of skin cancer.

Disease and diagnosis

1 Treat patient's DNA with restriction enzymes; separate DNA fragments by gel electrophoresis; pick up samples on a nylon membrane; add radioactive DNA probe; DNA probe is single strand with a base sequence complementary to a defective gene; binds if defective gene present and made visible by 'fogging' of X-ray film.

2 Test blood for amylase and trypsin; should not be in blood – if present, pancreatitis.

3 a Glucose oxidase acts on glucose, forming gluconic acid and hydrogen peroxide; peroxidase acts on peroxidase and blue chromagen dye – changing dye colour to green through to brown, depending on concentration of glucose; compare colour against colour chart to find glucose concentration.

  b Test for amylase – add starch/glycogen and look for formation of maltose/glucose.

Drugs in the control and treatment of disease

1 Shape of blocker allows it to bind to noradrenaline beta receptors; this prevents noradrenaline binding; stops heart muscles beating harder and prevents hypertension.

2 Bacteria cannot reproduce to a population size that is dangerous to the patient; after a few days, the patient's own immune system produces antibody that kills the bacteria and they recover.

3 a Identical antibodies produced by a clone of a single B-lymphocyte; highly specific to one antigen.

  b e.g. magic bullet – monoclonal antibody specific to antigen found on cancer cells but not normal cells; cytotoxic drug attached to antibody kills cancer cells.

Answers to end-of-module questions

Module 1: Molecules, Cells and Systems B   HB

1 a i Cellulose molecules are very long polysaccharides – their size makes them insoluble (and unreactive); being straight they lie closely side by side and bind by many hydrogen bonds – form strong fibres.

ii Starch molecules are very long, coiled and branched polysaccharides – can be broken down to glucose for respiration; large – insoluble, not affecting water potentials, unable to cross cell membranes and stay in the cell; branched – a compact store of chemical energy.

b

c Primary structure – order of amino acids in a polypeptide – determines whether the polypeptide forms a helix or folds on itself – secondary structure; where the chain folds and where bonds form between different parts of the chain gives a specific shape to the protein/polypeptide – tertiary structure.

2 a i Hydrolyse with acid, neutralise with alkali, test with Benedict's reagent, look for red precipitate.

ii Biuret test/copper sulphate solution and hydroxide, lilac colour is a positive result.

iii Add water and ethanol and shake – formation of an emulsion is a positive result.

b Digest protein with acid/protease, carry out chromatography, use two-way chromatography to separate amino acids that overlap with the first solvent, identify amino acids by R_f values.

3 a i Plant cell has a cell wall, a large vacuole and chloroplasts, animal cell does not.

ii Three of the following: cell membrane – controls movement of substances in and out of the cell; nucleus – contains the genetic information of the cell; mitochondrion – ATP is made in aerobic respiration; ribosomes – where proteins/polypeptides are made; rough ER – where proteins are made in the space of the ER; smooth ER – transports proteins made by the RER; Golgi body – packages proteins for transport round/out of the cell.

b Break open cells in ice-cold buffer; differential centrifugation to isolate cell fractions; electron microscopy to identify fraction with mitochondria; test mitochondrial fraction for its biochemical functions.

4 a i Arrows pointing from −350 kPa to −375 kPa and −400 kPa cells; also from −375 kPa to −400 kPa.

ii The more negative the water potential, the lower the concentration of water in the cell; water diffuses along concentration gradient from higher to lower; osmosis across the selectively permeable cell membranes.

b i Chloride ions moved against concentration gradient by active transport.

ii May be different numbers of carrier proteins for each ion; chloride ions might balance several types of positive ions.

iii Less oxygen – lower rate of respiration; less ATP for active transport; so internal concentrations of ions may fall.

5 a i Movement of ions/molecules randomly from higher to a lower concentration until evenly distributed.

ii Movement of ions/molecules by diffusion along a concentration gradient, through a membrane carrier or channel protein.

iii Movement of ions/molecules against a concentration gradient, using ATP from respiration.

b Water is a small, non-polar molecule which can pass through the phospholipid bilayer; glucose is much larger and has many polar groups, so cannot pass through bilayer – carrier protein binds to and carries glucose across.

6 a Enzymes are very specific; react with a particular substrate; lower the activation energy of chemical reactions; allow reactions to occur at the temperatures found in living organisms.

b i To start with, rate increases with the concentration of substrate, because more active sites are occupied as the substrate concentration increases; graph levels off when all of active sites are occupied most of the time.

ii The inhibitor lowers the maximum rate of action of the enzyme; it is a non-competitive inhibitor; binding to a site other than the active site, changing the shape of the enzyme and the active site.

7 a 45°C.

b i Low kinetic energy, so few collisions between enzyme and substrate.

ii Enzyme denatured; tertiary structure changed; enzyme no longer fits active site.

c i Non-boiled becomes active as kinetic energy increases; boiled no activity, because denatured.

ii pH of the solution – at optimum; concentration of substrate – higher produces higher rate; concentration of enzyme – if higher produces higher rate.

8 a Sinoatrial node in wall of right atrium; produces regular bursts of electrical impulses; these spread rapidly through walls of both atria – they contract together; impulses reach the atrioventricular node; delay of 0.15 seconds – so that ventricles contract after atria; impulses from AV node through bundle of His and branches to all parts of the ventricles; ventricles contract from bottom to top, pushing blood up and out into the arteries.

b i During exercise it needs to get more blood to muscles; carrying oxygen and glucose for faster respiration.

ii Rise in carbon dioxide in blood lowers pH and stimulates chemoreceptors; carotid/aortic sinus; nerve impulses to cardioaccelerator centre; in medulla; nerve impulses to SAN.

9 a Atrioventricular valves prevent backflow of blood from ventricle into atrium; semilunar valves prevent backflow of blood into ventricles from arteries.

b i X – semilunar valve opens because ventricle contracts – pressure inside becomes higher than in the aorta. Y – semilunar valve closes because ventricle is relaxing – pressure inside falls below that in aorta.

ii Time between corresponding points on the graph = 0.8 seconds, so rate = 60 ÷ 0.8 = 75 beats per minute.

Module 2: Making Use of Biology **B**

1 a Easier to isolate from cells. / No need to break cells open, or enzymes secreted into external medium. / Often secreted on their own. / Purification/ downstream processing easier/cheaper. / More pH- or temperature-stable.

b Pectinase (from *Aspergillus niger*); filter to remove fungus; evaporate to concentrate enzyme; precipitate pectinase as a solid and filter off; dry and grind to a powder.

2 a Very specific, so identify a specific substance they bind to/react with; high sensitivity, so react even with low concentration of substrate.

b i Makes it more stable, so less sensitive to environmental changes/temperature changes. / More likely to react with glucose as it crosses the membrane, so more accurate reading.

ii Glucose oxidase reacts with glucose to form gluconic acid; reaction uses oxygen; more glucose present, the more oxygen is used and this is measured by the probe.

3 a i Rice has aerenchyma, air spaces in the stem and roots; allows oxygen to diffuse to submerged roots for respiration.

ii sorghum has a thick waxy cuticle on its leaves; reduces water loss by evaporation from the surface of leaves; OR sorghum has stomata in sunken pits; reduces air movement past the stoma, water vapour accumulates just outside stoma and this reduces water potential gradient for diffusion of water out of stomata.

b Maize has a different enzyme/PEP carboxylase for 'fixing' carbon dioxide in photosynthesis; this has a very high affinity for CO_2 and allows efficient photosynthesis at low CO_2 concentrations; high temperature in tropics means stomata often closed to reduce water loss; this leads to low CO_2 inside leaves, which maize can cope with.

4 a Some pesticides are not broken down in organisms or excreted; this means that they accumulate in the tissues of the organism; if it is eaten, the next organism in the food chain accumulates the pesticide in its tissues – together with pesticide from all the other organisms it eats; the concentration of accumulated pesticide increases along a food chain.

b i Pigeon and moorhen lower pesticide content than sparrowhawk and heron; heron much higher than sparrowhawk; one mark for use of comparative figures from the graph.

ii Carnivores higher in their food chains; so greater bioaccumulation.

iii Longer food chain up to the heron; so more stages for bioaccumulation; OR fish accumulate pesticide more than birds or small mammals; so less bioaccumulation in sparrowhawk food chain.

5 a On a dull day light intensity is the limiting factor for photosynthesis; higher temperature and CO_2 from gas heaters will have no effect; so a waste of fuel, leading to lower profits.

b Large crop yield means a lot of mineral ions removed with crop; need to replace mineral ions in soil to maintain yield; especially nitrate, phosphate and potassium.

6 a Extra nitrate and phosphate in water causes rapid growth of algae; these smother and kill other plants which are also producers for food chains/webs; as algae die in large numbers, leads to rapid growth in population of decomposing bacteria; these remove oxygen from the water for respiration, causing suffocation of many fish and invertebrates.

b i Increase in nitrogen in large water plants and fish; greatest increase in large water plants; decrease in nitrogen in algae.

ii Reduced input of nitrogen/nitrate reduced eutrophication; this reduced algae population, allowing large water plants to grow; which form base of food chains for fish; fewer algae mean fewer bacteria; this leads to more oxygen in water for fish.

7 a Polymerase chain reaction; separate DNA strands by heating to 95°C; cool to 60–70°C to allow primers to attach to strands; DNA polymerase adds DNA nucleotides to produce new complementary strands to DNA; repeat cycle of heating and cooling, doubling DNA each time.

b i Restriction endonucleases used to produce DNA fragments; each enzyme cuts DNA at a specific base sequence.

ii Radioactive DNA probes; single strand DNA with a base sequence complementary to a particular VNTR; binds to its VNTR by complementary base pairing; made visible by 'fogging' of X-ray film.

c Person A, because shares same bands/VNTRs with blood sample.

8 a i Transcription.

ii Six-codon CGU appears twice.

iii GCAT.

b i Anticodon.

ii tRNA is 'clover leaf' in shape. tRNA standard length, mRNA is variable. tRNA has an amino acid binding site. tRNA has anticodon, mRNA has codons. tRNA has hydrogen bonds between base pairs.

iii Carries specific amino acid to ribosome. Anticodon attaches to codon on mRNA; leaves amino acid at ribosome and picks up same type of amino acid from cytoplasm.

9 a (i) 22. (ii) 44.

b Chromosomes carrying genes that control the same characteristic. The chromosomes are the same size.

c Meiosis reduces the diploid number to the haploid number. Gametes are haploid. Fusion of haploid gametes at fertilisation restores the diploid number.

d Splitting immature embryo provides identical cells. Process repeated several times. Developing clones are implanted into surrogate mothers.

10 a i Use of restriction endonuclease. Enzyme cuts DNA at specific base sequences.

ii Use of a vector e.g. plasmid. Cut plasmid with endonuclease. Join gene to DNA using ligase enzyme.

b Low temperature in North Sea. Slows down growth / activity / enzyme action of bacteria.

c Use of plasmid containing gene marker, e.g. for antibiotic resistance. Marker gene will enter the 'super Pseudomonas' with foreign gene. Culture bacteria on plate with antibiotic. Only 'super Pseudomonas' will grow.

Module 3: Pathogens and Disease

1 a i Restriction endonucleases used to cut DNA into fragments; each cuts DNA at specific base sequence.

ii Apply electrical potential; fragments of DNA travel towards the anode; smaller the fragment, the further/faster it travels.

b DNA probe single stranded; has base sequence complementary to specific gene.

c Child inherits DNA from both parents; so inherits different VNTRs from both – where restriction enzymes cut.

d Mother – since she has corresponding band.

2 X – RNA; carries genetic information.

Y – reverse transcriptase; produces HIV DNA from RNA.

Z – envelope; fuses with host cell plasma membrane, so virus can enter cell.

3 a Damaged endothelial cells; stop releasing anti-clotting factor; connective tissue exposed to the blood; platelets stick to damaged area and each other; all lead to production of active clotting factors and thromboplastin.

b On dashed arrow below factor X.

c i Thrombin

ii Converted to insoluble fibrous protein, fibrin, which forms a mesh to trap platelets and red cells.

d Inability of the blood to clot/a form of haemophilia.

4 a *Plasmodium* – e.g. intermediate mosquito host; to carry it from one human host to the next; asexual

reproduction/many cell divisions in host cells, to produce many offspring and ensure some get to new hosts.

b *Schistosoma* – e.g. intermediate snail host in which vast numbers of cercariae produced to ensure some infect new human host; in humans, larvae bore into gut or bladder, to ensure they are released to infect snails.

5 a i Formation of plaque in the wall of arteries; starts as fatty deposits; may harden and narrow lumen of artery, restricting blood flow.

ii Diet rich in animal/saturated fats/cholesterol – increases deposits of cholesterol in artery walls; smoking – nicotine causes constriction of arteries; causes high blood pressure damaging to artery walls; and increases chance of plaque formation.

b i A swelling of an artery where the wall is weakened.

ii Atheroma restricts blood flow, which leads to increased pressure (locally); this pressure can weaken the wall and lead to aneurysm; rough area of plaque (of atheroma) can cause blood clotting reaction; blood clot/thrombosis formed often sticks where lumen of artery is narrow due to atheroma, and stops blood flow.

c Stops blood supply to an area of heart muscle; muscle cells start to die, because of lack of oxygen and glucose for respiration; can result in heart attack.

6 a Normal cells' division controlled by genes; for growth factors that stimulate mitosis, and their plasma membrane receptors, mutation can produce abnormal genes/oncogenes that lead to uncontrolled cell division; mutation of tumour suppressor genes can also lead to uncontrolled division; no longer produce protein that can block growth factor.

b Benign tumour does not spread – cells stay together as a lump; malignant tumours metastasise; cells break away from the primary tumour and invade other organs, to form secondaries.

7 a Increases or decreases in smoking are followed by increase or decrease in lung cancer; figures to support statement; time lag between change in smoking and change in lung cancer.

b Smoke contains tar; tar contains carcinogens; the greater the amount of cigarettes smoked, the greater the intake of carcinogens, the greater the risk of developing lung cancer; carcinogens cause mutations of genes controlling cell division.

Index